Agent AI for Entrepreneurs

Automate the Busywork, Reclaim Your Time, & Build a Business That Runs Itself Without Losing Your Mind or Your Weekend

Roger Best

This publication is designed to provide accurate and authoritative information in regard to the subject matter covered. It is sold with the understanding that neither the author nor the publisher is engaged in rendering legal, investment, accounting or other professional services. While the publisher and author have used their best efforts in preparing this book, they make no representations or warranties with respect to the accuracy or completeness of the contents of this book and specifically disclaim any implied warranties of merchantability or fitness for a particular purpose. No warranty may be created or extended by sales representatives or written sales materials. The advice and strategies contained herein may not be suitable for your situation. You should consult with a professional when appropriate. Neither the publisher nor the author shall be liable for any loss of profit or any other commercial damages, including but not limited to special, incidental, consequential, personal, or other damages.

Contents

Introduction: The High Cost of Staying Stuck 1

1. Demystifying AI for Entrepreneurs 5
2. Strategic AI Integration 19
3. AI Applications Across Industries 35
4. Real-World Success Stories 49
5. Avoiding the AI Backfire 61
6. Tackling Time Constraints 73
7. Enhancing Business Efficiency with AI 85
8. Future Trends & Emerging Technologies 97
9. Building a Scalable AI Strategy 107
10. Creating a Culture of Continuous AI Learning 117

References 129

About the Author 135

Also by Roger Best 137

Introduction: The High Cost of Staying Stuck

Back in 2019, a simpler time when sourdough bread was still just bread and not a national coping strategy, a study revealed something truly sobering: the average entrepreneur spends nearly 70% of their time on what researchers gently called "busywork."

Not innovation. Not bold moves or inspired vision. Just the unrelenting parade of scheduling snafus, follow-up emails, customer queries, invoice juggling, and digital whack-a-mole that clogs your brain and calendar.

If that stat doesn't make you want to lie down with a cold cloth and reconsider your life choices, I don't know what will.

Now, what if you could take all that time, all those hours eaten alive by admin tasks and tactical chaos, and pour it back into what's important to *you*? Into strategy, into growth, into finally drinking a hot cup of coffee before it goes the way of many childhood dreams: lukewarm and abandoned.

If you're reading this, I'm guessing you're an entrepreneur. Maybe a business owner in the 35-to-65-year-old camp who built something real, from the ground up. You've worn every hat in the building, sometimes all at once, visionary in the morning, customer service rep by lunch, and head custodian by close of business. You don't need me to explain "overwhelmed." You *are* the definition.

I've been there too. I once found myself late on a Tuesday afternoon, standing in my kitchen holding a sticky note in one hand and a half-eaten protein bar in the other, trying to remember what day it was, while still considering the email I just walked away from that was marked "URGENT" ... it was about printer ink. This is the entrepreneurial condition: a dream that somehow got buried under an avalanche of checklists and constant interruptions.

But here's the good news: it doesn't have to stay this way.

And no, I'm not about to introduce you to your new robot overlord. No creepy voice assistants trying to sound like Morgan Freeman. No sentient machines plotting to steal your job *or* your dog.

What I'm talking about is much simpler and far more helpful.

Meet Agent AI. Your digital sidekick. Your behind-the-scenes wizard. Your infinitely patient intern who never needs sleep, doesn't complain, and will never eat your snacks.

Agent AI isn't about replacing people. It's about *freeing* them. AI, and AI Agents specialize in the dull stuff, scheduling, follow-ups, customer responses, research, and reporting. The things that chip away at your time and your sanity like a woodpecker on Red Bull.

And here's the beauty of it: while your agent handles all that, you can get back to what actually matters. Creating. Growing. Thinking clearly. Or maybe, just maybe, having a Wednesday afternoon all to yourself for the first time in years.

This book is your guide to that new reality.

You don't need to be a coder, a tech junkie, or someone who gets excited about neural networks. You just need to be someone who is sick of being buried in busy work and ready to take back control of your time. We'll cover the basics of what AI agents are, how they work, and, most importantly, how you can start using them in your business *this week*, not next decade.

We'll look at real-world use cases: customer service automation, more innovative marketing, scheduling magic, and even how AI can help you make better decisions faster. And because you're a responsible human being (the kind who wants your business to grow without becoming morally questionable), we'll also talk about using these tools ethically, inclusively, and without accidentally creeping out your customers.

Each chapter in this book lays the groundwork; clear, straightforward, no-nonsense insights into what Agentic AI can actually do for your business (and what it can't). There are no worksheets, checkboxes, or brain-draining exercises here, those come later. This book is about helping you get your bearings before you sprint off in twenty different tech directions. Think of it as a map, not the mountain climb. A guide to help you avoid the rabbit holes and shiny distractions, so when you're ready to take action, you'll know exactly where to start, and how not to waste your time or budget.

So what's in it for you? More time. Less chaos. A business that hums along while you're hiking a mountain, grilling in the backyard, or simply staring at a wall because you finally have the luxury of being bored again. You get to scale without sacrificing your soul, or your sanity.

If that sounds like the kind of freedom you've been chasing since day one, then you're in the right place.

You've built something incredible. Now let's make it sustainable.

Ready? Let's begin.

Chapter One

Demystifying AI for Entrepreneurs

Or, "How I Learned to Stop Worrying and Love the Robot"

If you've ever woken up, glanced at your calendar, and thought, "Ah, yes, another day of trying not to drown in my own to-do list," then welcome. You're not just in the right place, you're among friends.

Entrepreneurship, as it turns out, is less about sipping cocktails on the beach and more about juggling flaming swords while balancing on a yoga ball. One minute you're strategizing your next big move; the next, you're deep in an email chain about printer ink. Somewhere along the line, the freedom you were promised started looking suspiciously like a hamster wheel with better Wi-Fi.

But what if there were a way to clone your most productive self, not in a sci-fi "beam me up" sort of way, but in a practical, sensible, dare I say liberating way? A version of you that handles the grunt work, while the real you gets back to building the life and business you actually wanted in the first place.

That's where AI agents come in. No glowing red eyes. No plans for world domination. Just tireless, invisible helpers that quietly take the busy work off your plate. Think of them as digital interns who never sleep, don't need coffee, and won't accidentally RSVP you to a toddler's birthday party.

Understanding AI Agents: The Basics, Without the Buzzwords

Let's get one thing straight right off the bat: AI agents are not here to replace you, outshine you, or take over the family business. They don't want your job, your glory, or your frequent flyer miles. They just want to help.

At their core, AI agents are like digital Swiss Army knives, except instead of blades and bottle openers, they come packed with algorithms and the uncanny ability to make sense of chaos. These aren't your chatty phone assistants or those slightly nosy smart speakers. No, AI agents are built to do real work. We're talking lead tracking, email follow-up, appointment setting, task reminders, even handling onboarding or sending project updates, all quietly running in the background. At the same time, you focus on the big stuff (or maybe eat lunch without your laptop).

And the best part? They don't need sleep, snacks, or supervision. They show up, do the job, and keep going, like the most efficient team member you've ever had to hire.

This isn't about science fiction or bleeding-edge tech. It's about finally having the kind of support system that doesn't call in sick or ask you how to open a PDF. It's about creating breathing room in your day without cloning yourself or building an empire of minions.

And no, you don't need a PhD to use them. You just need to know where they fit in, and that's exactly what we're going to figure out together.

Why Entrepreneurs Need AI Agents Now More Than Ever

Let's face it, entrepreneurship was supposed to be the great escape. You didn't leave your 9-to-5 to work 9-to-midnight, only now without paid holidays or a functioning HR department. You started this journey for freedom, flexibility, maybe even to have Fridays off just because. But somewhere along the way, the dream got hijacked by a never-ending parade of tiny tasks, inbox avalanches, and to-do lists that reproduce like fruit flies on caffeine.

If you've ever looked at your calendar and wondered, "How am I this busy and still not getting the important stuff done?" Congratulations, you're in excellent company.

Modern entrepreneurship often feels less like running a business and more like being chased by one. There's marketing to do, clients to follow up with, invoices to send, systems to fix, team members to train, fires to put out, and don't forget the brilliant new idea you had last Tuesday but haven't touched since because... well, you were busy deleting 37 spam emails and checking if your Stripe payment went through. Yes, AI Agents can do that stuff for you.

This is precisely why AI agents matter: they're built to take that relentless swirl of busyness and start untangling it. They free up your time not by giving you more to do, but by quietly doing the stuff you shouldn't have to do in the first place. No burnout. No drama. Just help where you need it most.

In a world that keeps telling entrepreneurs to hustle harder, AI agents offer a more innovative way forward, one where you actually get to enjoy the business you built.

What Makes AI Agents Different from Traditional Automation Tools

If you've ever used automation tools before, email sequences, calendar schedulers, or those neat little rules in your inbox, you might be thinking, "What's the big deal with AI agents?" And that's fair. At first glance, they may sound like fancy new versions of things we've seen before. But here's the difference: AI agents don't just follow rules, they learn.

Think of traditional automation as your trusty slow cooker. It'll do the job, but only if you follow the recipe precisely. AI agents, on the other hand, are more like that magical kitchen wizard who watches you cook once and then takes over while you sip something on the patio. They observe your routines, learn your preferences, and even adapt when the situation shifts.

Where old-school tools say, "If A happens, do B," AI agents say, "Hmm, A usually leads to B, but today feels more like a B-minus with a side of C." That's where the magic happens; they don't just execute; they interpret.

Need someone (or something) to prioritize emails, tailor responses, or notice when your customer's tone shifts from curious to cranky? An AI agent can spot the nuance. These agents aren't just task doers, they're intelligent collaborators, working quietly in the background to keep your business humming along while you finally get to focus on the big-picture stuff.

This is more than efficiency; it's a step toward sanity.

Where They Fit In (and Why Your Calendar Will Thank You)

Let's get practical. What can AI agents *actually* do in your business?

- **Chatbots**: Tireless little greeters who manage customer inquiries 24/7 without eye rolls or typos.

- **Virtual Assistants**: They sort your email, set your appointments, send reminders, and basically make sure you don't miss your dentist appointment (again).

- **Smart Analyzers**: These crunch vast oceans of data in moments, spotting patterns, trends, and oddities that would take you a team of analysts and at least three whiteboards to decipher.

They're designed to handle the boring bits so you can focus on strategy, growth, and, dare I say, even your personal life.

Think about what it would feel like to spend less time entering data and more time *interpreting* it. Less time scheduling meetings, more time closing deals. AI agents don't just help; they buy back time for you ... actual, usable hours.

Quick Pause: A Little Reflection

Here's a simple but powerful exercise: Take out a notepad or open that trusty Notes app and jot down five tasks that eat up your day but don't contribute meaningfully to your business growth. Things like sorting your inbox, updating spreadsheets, and chasing down meeting confirmations. Now ask yourself: "Could an AI agent handle this for me?"

Chances are that the answer is yes. And once you've offloaded those tasks, you'll wonder how you ever lived without them.

A Handy Little Glossary (Because We're Nice Like That)

Let's pin these terms to the bulletin board for reference (because I don't want to bore you with the techno-babble), and just in case you want to look them up for more details:

- **Machine Learning** – Computers learning from data to make better predictions. Like a dog, but more into math.

- **Neural Networks** – AI's version of your brain, built to recognize patterns. Less emotional baggage.

- **Algorithms** – Digital recipes. Follow steps, get predictable results (unless it's banana bread, which is always unpredictable).

- **Data Processing** – Turning raw info into usable insight. Like panning for gold, but with customer data.

- **Training Models** – Feeding data to AI so it gets better at its job. Like rehearsal, but without the jazz hands.

Still Feeling Overwhelmed? You Don't Have to Be

Here's the truth: AI goes far beyond understanding every technical detail. It can take you to a place where you know enough to see what's possible and pick the right tools to help your business do more, faster. You don't need to become a tech wizard. You need to stop fearing the curtain and take a peek behind the scenes.

With a bit of guidance, these tools will feel less like magic and more like what they really are: time-savers, insight-generators, and glorified digital interns that never forget a task.

And let's be honest: if something can cut your workload, sharpen your strategy, and help you get home before dark, it deserves a place in your business toolbox, even if you never learn how to spell "neural network" without looking it up.

So, let's keep going. There's more AI magic to uncover, and I promise we'll do it without once uttering the phrase "quantum computing."

Probably.

The Practical Entrepreneur's Guide to AI Concepts

(*Now with 100% more clarity and 100% less techno-babble*)

Let's be honest: if you've ever tried to wade through the murky waters of AI terminology online, you've likely come away feeling as if you've accidentally enrolled in a graduate-level computer science class. Words like "machine learning," "neural networks," and "algorithmic bias" get tossed around with the breezy overconfidence of people who clearly aren't worried about sounding confusing. So, let's take a moment to *actually* unpack what matters, with no jargon, no math, and no headaches.

What Exactly *Is* an AI Agent?

At its core, **an AI agent** is a digital assistant that doesn't just sit there waiting to be told what to do; it observes, thinks (or, more accurately, approximates thinking), and acts.

Let's say that you're hiring a new employee named Alex. You grant Alex access to your inbox, customer relationship management (CRM) system, calendar, and sales data. Now, unlike a normal assistant, Alex doesn't come to you every five minutes asking what to do next. Instead, they learn. They notice that customers who haven't been contacted in three days are less likely to buy, so they automatically schedule follow-ups. They realize your team meetings tend to run overtime on Wednesdays, so they start blocking off a buffer in your calendar. They even detect that a specific kind of lead responds best when approached via text instead of email, and they adjust accordingly.

That's an **AI agent** in action.

To put it plainly, an AI agent is a type of software that does the following:

- Takes in information from its environment (like incoming emails or project updates),
- Decides what needs to be done based on a set of goals you've defined,
- And acts on your behalf, without you needing to manually push the buttons.

Unlike basic automation (which performs the same task repeatedly, much like a very obedient toaster), an AI agent can adjust its behavior based on what it learns. It's not sentient, mind you; no inner monologue or existential crisis is going on, but it's responsive, adaptive, and often eerily good at taking the "ugh" out of your daily grind.

A Clear Example: Meet "Follow-Up Fran"

Let's say you run a coaching business, and you often forget to follow up with leads who've downloaded your free lead magnet. They slip through the cracks while you're neck-deep in a proposal or trying to fix the office printer (again).

You install a lightweight AI agent named "Follow-Up Fran." Here's what she does:

1. She monitors new lead activity through your CRM.

2. She waits 48 hours after a lead downloads your guide, then drafts a personalized follow-up email using the lead's name, interests, and past interactions.

3. If the lead replies, Fran forwards the message to you and marks them as "engaged."

4. If the lead *doesn't* reply, Fran follows up again after three days with a short message and a link to schedule a call.

5. She adjusts the tone and timing of her follow-ups based on what works best over time.

And the best part? She does all of this quietly, efficiently, and without you lifting a finger.

That's not just automation; that's Agentic AI, which is the next level of productivity. It doesn't just follow instructions; it makes choices based on outcomes to help you reach your goals.

The takeaway? You don't need a robotics lab or a PhD to start using AI agents in your business. What you *do* need is a clear idea of what you want help with, and a willingness to let AI do the boring parts, so you can focus on the big, exciting, *actually-human* parts of being an entrepreneur.

Now that we've defined our digital co-pilot, you're ready to meet the many ways these agents can swoop in and transform the way you run your business. As we move into the next chapters, we'll show you exactly how to assess your business for AI-readiness, choose the right tools, and roll things out without turning your office into a scene from a sci-fi movie.

Spoiler alert: It's easier (and more fun) than you think.

Implementing AI Without Losing Sleep (or Sanity)

Now, if you're thinking, "This all sounds great, but where do I even begin?", you're not alone. The truth is, integrating AI into your workflow can be as simple or as complex as you want it to be.

Start small. Identify the repetitive tasks that quietly steal your time each week. Think data entry, follow-up emails, scheduling, reports, etc., and choose one. Just one. Then, find an AI tool that handles it well. Don't worry about overhauling your entire system overnight. Begin with something manageable. Implement it. Watch it work. Get feedback. Tweak. Expand. Before you know it, you'll be wondering how you ever got by without it.

And yes, this applies to project management too. AI doesn't just babysit your to-do list; it can assign tasks, track deadlines, and gently remind team members when things are due. It spots bottlenecks before they become full-blown emergencies and ensures projects move along smoothly. Like having a polite, all-seeing office manager with an exceptional memory and no need for coffee breaks.

Real-World Examples (Without the Buzzwords)

Let's move past theory and peek into the real world, where people are already using AI to solve actual business problems:

Take an e-commerce store that started using AI to recommend products based on customer behavior. Suddenly, shoppers were seeing exactly what they wanted (did you see Minority Report), and sales soared. No magic. Just good use of data.

Or consider the accounting firm that automated its routine financial reports. Human errors decreased, time savings soared, and the team was able to focus on more strategic work, like figuring out how to utilize their newfound free time.

And here's the important part: these weren't massive corporations with limitless tech budgets. These were normal-sized businesses doing very clever things, one step at a time.

But What About the Cost? And the Tech Stuff?

Good questions. The idea that AI is wildly expensive is, thankfully, outdated. Many platforms are surprisingly affordable and built with small businesses in mind. You can start with a pilot project, dip your toe in the water before cannonballing in. There's no need to buy the whole theme park.

As for the complexity? That's also changing. User-friendly AI tools are now designed to be more intuitive than ever. No coding required. Most come with tutorials, support teams, and communities of fellow users who've already tripped over the same hurdles and lived to tell the tale.

If you've ever figured out how to send a newsletter, set up a website, or troubleshoot a printer, congratulations, you already have the skills to work with AI tools. It's just a matter of taking those same "figure-it-out" muscles and stretching them a little further.

This Isn't Hype. This Is Help.

Let's be honest, AI gets tossed around like confetti in business articles these days. But underneath the buzzword is a very practical reality: it helps you work smarter. It saves time. It cuts waste. It makes your team more efficient. And it lets you focus on what actually matters: strategy, growth, creativity, and maybe even a proper lunch break.

You don't need to "keep up" with tech for the sake of it. You need to keep up because the tools now available are too helpful to ignore. Integrate AI into your operations, and you're not just reacting to change; you're driving it.

So go on, explore a little. Start small. Tinker. Experiment. You don't have to become an AI expert overnight. You just have to be willing to take that first step.

And who knows? By this time next month, you might be sipping coffee while your AI agent finishes your reporting, organizes your inbox, and sends out your follow-ups, all before you've had your second pastry.

Which, if we're being honest, is exactly the kind of future we've all been hoping for.

AI Myths Debunked: Separating Hype from Reality

(Also: Why AI Isn't Coming for Your Latte Machine)

Let's be honest. AI has suffered from a case of over-hype. One minute it's curing diseases and writing novels. The next it's supposedly running Fortune 500 companies and hosting late-night talk shows. Some headlines make it sound like we're all about two weeks away from being politely asked to step aside by a sentient spreadsheet.

But reality? Much more down-to-earth.

AI is, at its core, a very advanced calculator with a better grasp of grammar. It can analyze data, spot trends, and help you make faster, more informed decisions. But it's not psychic, omnipotent, or even all that clever when taken out of context.

Remember the AI that was hyped to diagnose medical conditions better than doctors? Turned out it was a bit less accurate than your average GP after a good night's sleep and a half-decent sandwich. It's powerful, yes. But it still needs supervision, structure, and real-world context to be useful.

So, What *Is* AI Actually Good At?

- **Customer segmentation**: Finding patterns in buyer behavior and grouping people accordingly. It's like high school cliques, but less judgmental.

- **Forecasting demand**: Spotting seasonal trends so you don't overstock Halloween decorations in March.

- **Streamlining operations**: Optimizing everything from inventory to internal communications.

- **Personalizing experiences**: Recommending just the right thing to the right person, without being creepy. Ideally.

It's not world domination. It's helpful, practical, behind-the-scenes support.

And What Isn't It Good At?

Well... human emotions, for starters. AI might detect "frustration" in a customer's message, but it still doesn't really *get* why your client is annoyed that you sent them a "Happy Anniversary!" coupon when they're newly single.

It also struggles with the unexpected. Give it a banana-shaped package in a mail-sorting algorithm, and suddenly your digital genius is having a small existential crisis. It's great at what it knows, but still needs you for the weird stuff. (Which, in business, is most days.)

What This Book Will Teach You (Without Melting Your Brain)

This book is your practical, plain-English guide to making AI agents work for you, not the other way around. You won't need to learn to code, memorize acronyms, or grow a hoodie and suddenly start referring to your garage as a "lab." If you've ever looked at a new piece of software and thought, "Well, there goes my weekend," rest assured: this isn't that.

Instead, you'll get the kind of advice that makes sense over your morning coffee. We'll cover what AI agents actually are (and what they're not), how to choose the right option, and how to integrate them into your day-to-day operations in a way that saves you time, money, and mental bandwidth.

This book was written with entrepreneurs like you in mind, especially those between 35 and 65 who've built real businesses and want to make smart, sustainable improvements. You're not chasing shiny objects. You're chasing freedom. And we're going to help you reclaim it.

You'll learn how to delegate tasks to AI agents without losing your personal touch, how to automate customer service without sounding robotic, and how to run your business more efficiently while still keeping it human.

Why This Isn't Just Another Tech Hype Book

Let's be honest, there's no shortage of books out there promising to revolutionize your business with a dash of artificial intelligence and a sprinkle of buzzwords. This isn't one of them.

What sets this book apart is its commitment to simplicity and sanity. We'll start at the actual beginning, not somewhere halfway through a TED Talk. You'll get a clear, no-nonsense breakdown of what AI agents are and how they actually work, not what they might someday become in a lab.

Then we'll ease into how to use them. Not all at once. Not by next Friday. But in small, manageable steps. You'll learn how to automate things like customer service, marketing outreach, scheduling, and even decision-making, without hiring a tech team or sacrificing your business's soul to the machine.

And because we're grown-ups here and not characters in a science fiction movie, we'll also talk about the serious stuff: ethics, privacy, bias, and the line between helpful assistants and creepy algorithmic overreach. No doom, no drama, just what you need to know to use AI responsibly and confidently.

How to Use This Book (and Not Let It Collect Digital Dust)

This book wasn't designed to impress your bookshelf. It's meant to be dog-eared, scribbled in, and, if I've done my job, mildly coffee-stained by the end.

No theory for theory's sake. No tech lectures. Just straight-up strategies and real-life examples you can use in your business, probably by Tuesday.

What You're Probably Thinking Right Now (And Why That's Perfectly Reasonable)

Let me guess, some part of you is thinking, "This all sounds great, but isn't AI a bit... much?" Maybe it feels like the kind of thing reserved for hoodie-wearing geniuses in tech incubators or companies with billion-dollar R&D budgets.

You're not wrong to be skeptical. AI can feel like it's either overhyped or terrifying, depending on the headline of the day. But here's the truth: it's neither. It's not magic, and it's not mayhem. It's a tool, one that's finally accessible to the rest of us.

No buzzwords. No smoke. No mirrors. Just straight-up, clear-headed guidance on how to use AI to make your business run better and your life a whole lot saner. And yes, we'll throw in a few laughs along the way, because if you're going to rethink how you run your business, it should at least be a good time.

So, What's in It for You?

You're out on a hike, birds chirping, sun on your face, not a single bar of cell service in sight, and somehow, miraculously, your business hasn't burst into flames. Emails are being answered. Leads are getting the attention they crave. Projects are humming along as if you were still chained to your desk, except you're not. You're miles away, possibly wondering if you packed enough trail mix. This isn't science fiction or some Silicon Valley fever dream. It's what AI agents can do. Not someday. Today.

What's in it for you is your time back. Your clarity back. Your weekends back. A business that works *for* you, not because of you. One that grows without dragging you down with it.

And you don't need to become a tech wizard to make it happen. Just a little curiosity and a willingness to try something new. You've already done the hardest part: you've built something real from the ground up. Now it's time to make it resilient, scalable, and a lot more peaceful.

Ready to trade the chaos for something a bit more clever?

Good. Let's get started.

Chapter Two

Strategic AI Integration

Or, Don't Just Plug It in and Hope for the Best!

LET'S SAY YOU'VE JUST unboxed the fanciest, flashiest espresso machine on the planet. Chrome finish. Touch screen. Claims it can speak Italian. But as you plug it in, you realize your kitchen's wiring hasn't been updated since disco was king, and the nearest outlet is currently occupied by a toaster that only works if you hold it shut with a wooden spoon.

That, in essence, is what implementing AI feels like when your business hasn't done a bit of prep work.

Before you start dreaming of dashboards doing all the thinking while you sip mojitos on a Tuesday, you need to know where you stand. And I get it, this part isn't glamorous. It's not the TED Talk moment. But it is, perhaps annoyingly, essential.

This is where you step into the role of business detective. What's clunky? What's glitchy? What turns normally pleasant people into snippy, spreadsheet-hurling goblins by 2:00 p.m.? Find the friction. Because under all that friction is an opportunity for AI to earn its keep.

And don't skip the human audit either. Are your people open to trying something new, or do they act like you're trying to introduce a robot uprising when you say the word "automation"? You don't need them to be data scientists. You just need them to be willing.

What's In It for You?Time saved. Energy reclaimed. Systems that make sense. And best of all? You stop being the bottleneck in your own business.

So grab a whiteboard, a marker, and perhaps a cup of tea (see, I can do more than coffee in my illustrations). We're going sleuthing.

Assessing Your Business for AI Readiness

Step One: Spot the Friction

Think like a detective, or better yet, a slightly nosy friend who opens your cupboards just to see what's inside. You're not looking to judge; you're looking to understand.

Where are the slowdowns? Where does the work pile up or get repeated unnecessarily? Is there a particular task or system that turns otherwise sane humans into muttering, eye-twitching creatures of dread? That's your cue.

Anything that feels like déjà vu, sending the same email for the 37th time, copying data from one system to another like a digital courier, deserves a closer look. These are your automation gold mines, just waiting to be tapped.

Key Takeaway: You can't improve what you haven't diagnosed. Your first mission is to uncover the bottlenecks, the redundancies, and the "Why are we still doing it this way?" moments in your business.

Step Two: Are You *Actually* Ready?

Let's talk infrastructure, not in a dull, IT-manual kind of way, but in a "Can your business handle this new horsepower without throwing a rod?" kind of way.

First up: tech readiness. If your current systems are hanging on by a thread (or still using software that sends regular pop-ups about Y2K compliance), AI might bring more frustration than innovation. You don't need cutting-edge tech, but you do need a stable foundation.

Then there's the human side. Are your people onboard? Curious? Or already picturing a robot uprising? The goal isn't to turn your team into data scientists, just to make sure they're not terrified of dashboards. Some light upskilling might be in order, think "AI 101," not "Advanced Machine Learning in Swahili."

Key Takeaway: Technical tools are only half the equation. If your systems and your team aren't ready for change, AI won't be a solution. It'll be a stress test.

Step Three: Know Your *Why* (And Don't Let It Be "Because It's Trendy")

Now that you've peeked under the hood, it's time to ask the big question: What problem are we actually trying to solve?

This is your "needs analysis," which sounds dreadfully corporate, but really just means: Are you chasing AI because it's shiny, or because it's useful?

Let's say your customer service reps are drowning in basic questions. That's a great candidate for a chatbot. Or maybe inventory keeps spiraling out of control, AI can help predict demand, spot trends, and reduce those awkward "we're out of stock again" moments.

Key Takeaway: Your AI implementation should be a solution to a real problem, not a vanity project. Anchor every decision to a business goal that matters.

Step Four: Assemble Your (AI) Avengers

Time to build your AI A-Team. And no, it's not just the IT folks who should be in the room. You need a cross-section of your business: the spreadsheet wizards, the operations gurus, the strategic thinkers, and yes, someone with access to the budget.

Think of it like a heist movie: you need the planner, the hacker, the inside man, and the guy who can charm the CFO into saying yes. This isn't a solo mission, it's collaborative, strategic, and very human.

Key Takeaway:AI adoption isn't just a tech decision. It's a business transformation. Get buy-in early from every part of your organization.

Bonus: The "Are We Sure?" Checklist

Before you even think about installing a single tool, run through this quick sanity check:

- **IT Readiness:** Do we have the infrastructure and support staff to manage new systems?
- **Process Awareness:** Do we understand our current workflows enough to know where AI fits?
- **Team Willingness:** Are our people onboard, curious, or going into witness protection at the mention of AI?
- **Executive Sponsorship:** Is someone in leadership willing to back this, budget it, and champion it?

If you've ticked most of those boxes, congratulations, you're ready to start building something transformative. If not, don't panic. It just means there's still some groundwork to do. Better now than halfway through a failed rollout.

Prep Before You Plug

Strategic AI integration starts with clarity. Clarity about your systems, your people, your pain points, and your purpose. Get those pieces aligned, and you won't just "add AI to your business", you'll amplify what already works and smooth out what doesn't.

And if that sounds like a lot, that's because it is. But it's also where the magic starts.

Next up? We turn that prep into a real plan, complete with pilot programs, feedback loops, and contingency plans that (hopefully) don't involve fire.

Developing a Step-by-Step AI Integration Plan

Picture yourself standing in front of a blank canvas, a palette full of colors in one hand, a brush in the other, and absolutely no idea where to start. This is what planning AI integration can feel like. So much possibility, so many options, and just enough uncertainty to make you want to pretend you didn't buy the canvas in the first place.

But fear not. We're not here to splatter data and hope for the best. We're here to turn that mess of color into something closer to a masterpiece, or at least a nice, functional flowchart.

Begin with a Plan (Before You Begin With AI)

First, let's talk objectives. Without clear goals, you're just poking buttons and hoping the future shows up. Think of your objectives and timelines as your GPS system. You don't have to know every bend in the road, but you *do* need to know what direction you're heading.

Set milestones and KPIs, the business world's version of mile markers and pit stops. These help you and your team know where you are, how far you've come, and whether you're still headed toward "Operational Efficiency Nirvana" or quietly detouring toward "Where Did All My Time Go-ville."

And don't just plan for the sunny days. Include review points and recalibration zones. You *will* hit the occasional pothole; maybe the AI tool won't integrate with your CRM, or maybe Dave from accounting still insists on using fax. That's fine. You're not looking for perfection. You're looking for progress.

Feedback Loops: Because Even the Best Plans Get Weird

There is no such thing as a flawless tech rollout. I don't care what the glossy brochure says. Somewhere, someone will forget a password, miss a training video, or try to use a voice assistant to microwave their lunch.

That's why feedback loops are essential. They're your built-in course correction mechanism. Regular team check-ins serve as both safety net and sanity check. What's working? What's not? What did we wildly underestimate? What's surprisingly easy?

Then, here's the magic, **you adapt**. You tweak. You refine. AI integration, like life, is rarely a straight line. Think of it more like wandering through IKEA with an Allen wrench and an optimistic smile.

Start Small: The Pilot Program (AKA Test-Driving the Future)

Before you unleash AI across the entire company like an enthusiastic golden retriever in a China shop, test it. Choose a department or workflow where the stakes are manageable but the potential for improvement is real.

This is your **pilot program**, the AI equivalent of dipping your toe in the pool to see if it's warm enough. Measure performance with the metrics you defined earlier. Was it faster? Was it smoother? Did anyone cry? These are all important benchmarks.

A successful pilot gives you confidence. A rocky one gives you insight. Either way, you win, because now you have data, and data is what separates hopeful guesses from smart decisions.

Build in Flexibility (Because Tech Is a Moving Target)

One thing about technology: just when you think you've got it figured out, it does a little twirl and changes outfits.

So make sure your AI integration plan isn't written in stone. Write it in pencil. With lots of erasers. Or better yet, make it a living document, able to grow, bend, and shift with the needs of your business and the whims of the software gods.

Scalability is key here. Roll it out department by department. Don't try to do everything at once, unless you enjoy caffeine-induced panic and late-night troubleshooting sessions. Gradual expansion allows each team to adjust, train, and provide input without setting anything on fire.

Also, **have a contingency plan**. If AI throws you a curveball, or the platform you love suddenly stops existing (it happens), you want to be prepared. Like keeping a poncho in your glovebox. You might never need it, but you'll be glad it's there if things get drizzly.

A Roadmap Worth Following

Here's a sample framework you can adapt to your liking. You may know, if you've read some of my other books, that I hate bullet points and checklists. And yes, this is a checklist, but I figure you'd rather see this short checklist than have to read the two pages of details if I wrote it all out.

Sample AI Integration Roadmap Template

- **Objective Setting** – What are we trying to fix, improve, or revolutionize?
- **Timeline Planning** – When should this happen, and how will we know we're on track?
- **Milestone Identification** – Break it down into snackable chunks.
- **Feedback Incorporation** – Bake in regular checkpoints for course correction.
- **Pilot Program Implementation** – Try before you buy (or before you break anything).
- **Scalability Strategies** – Expand slowly, support often.
- **Contingency Planning** – Have a Plan B, C, and maybe even a Plan "Unplug and Reboot."

Keep in mind, this plan is a roadmap, not a contract. Roads close. Detours happen. Don't be afraid to reroute if a better option appears. Some of the most innovative solutions come from veering off the original path and finding a shortcut you didn't know existed.

Celebrate as You Go

We don't do this enough in business, but we absolutely should: **celebrate small wins**.

Finished your pilot? Give your team a shoutout. Automated a clunky manual task? Share it in your team chat. Someone finally learned how to pronounce "algorithm" correctly? That's worth a donut.

Momentum is built on morale, and nothing keeps a team moving forward like feeling seen, appreciated, and caffeinated.

Bring People With You

One final note, possibly the most important of all: **AI isn't just about tech. It's about people**.

The best technology in the world won't work if the people using it aren't engaged. Involve your team from the beginning. Listen to their concerns. Ask for their ideas. Let them poke at the tools, question the process, and make suggestions. Not only does this foster buy-in, it often leads to better results, because no one knows your day-to-day pain points better than the people living them.

In Summary: A Little Chaos, a Lot of Potential

Integrating AI into your business is like switching from riding a bike to driving a car. It's faster, smoother, and comes with power windows, but you do need a plan, a license, and a little practice before you hit the freeway.

So grab your roadmap. Gather your people. Start small. Stay flexible. Celebrate often. And if something goes sideways? Don't worry. It's all part of the journey.

Coming up next: how to choose the *right* AI option for your business, without getting lured in by the ones that sound cool but secretly require three engineers, a translator, and a blood sacrifice to install.

Choosing the Right AI Tools for Your Business Needs

(Or: How Not to Buy a Gold-Plated Toaster When You Just Need Toast)

Imagine stepping up to a buffet so overwhelming, you start to question whether you're hungry or just mildly panicked. That's what choosing an AI tool can feel like. Sleek dashboards, dazzling claims, tools that promise to do everything short of walking your dog, it's all very exciting until you realize you have absolutely no idea which dish to put on your plate.

The key is to approach it like Sherlock Holmes on a double espresso: focused, skeptical, and ready to dig. The best place to start? Research. Look at what businesses like yours are using. Don't be dazzled by Fortune 500 case studies if you're running a 12-person operation with a perfectly decent espresso machine and a tendency to work out of coffee shops.

Check reviews, forums, and tech blogs, but do it with a critical eye. Some reviews are more curated than a Pinterest board. Look for recurring themes: Does the tool do what it promises? Does customer support respond faster than the speed of despair? Any mention of data loss, mysterious errors, or vanishing buttons? Red flags aren't always dramatic; sometimes, they're just too many people saying "It's fine" through gritted teeth.

Let's Talk Dollars (and Sense)

Once you've identified a few promising options, it's time for a bit of financial soul-searching.

Don't get lured in by the most expensive option simply because it comes with words like "quantum" or "enterprise-level." That's like buying a luxury yacht when all you really need is a rowboat that doesn't leak.

Instead, analyze the potential return on investment. Will this tool actually save you time? Will it reduce labor costs? Will it keep someone from spending their Monday mornings sorting data by hand while questioning their life choices?

Yes, automation can lead to huge savings. But weigh those gains against the price tag, setup costs, subscription fees, and the occasional hidden "integration service" that turns out to be a three-week engagement with a very serious consultant named Gary.

Think long-term, not just launch-day. The right tool should grow with your business, not bankrupt it while promising to optimize "synergy loops."

If It's Not Easy to Use, It's Useless

You know what's better than a powerful AI tool? One your team can actually use without sending you desperate messages at 11:47 p.m.

User-friendly interfaces matter. A lot. Prioritize platforms that are intuitive, such as those with drag-and-drop menus, visual dashboards, and tutorials that don't feel like they were translated from a Martian dialect. Your AI Agents should feel less like a plane cockpit and more like your favorite app: straightforward, forgiving, and dare I say, even pleasant to use.

The easier it is for your team to adopt, the faster you'll see results. And the fewer passive-aggressive comments you'll get in Slack.

Try Before You Buy (Because You're Not That Impulsive)

Don't commit to a new AI tool until you've taken it for a spin. Arrange a pilot or free trial, something that lets you kick the tires, test the fit, and maybe see how it performs under actual workload pressure.

During the trial, pay attention to how it feels in your workflow. Does it slot in smoothly, or is it the software equivalent of a squeaky shopping cart wheel? Does your team like it? Can they understand it without invoking the phrase, "I need a drink"?

This hands-on time is invaluable. It lets you assess performance, test compatibility with your existing systems, and gather genuine feedback from your team. And yes, their feedback matters. They're the ones who'll be using it every day, not the salesperson who called you "boss" three times in a single conversation.

Putting It All Together Without Losing Your Mind

Yes, the process of choosing AI tools can feel overwhelming. But by staying focused on your business goals, budget, and team capabilities, you'll make smarter decisions and avoid tools that look impressive but function like a blender with no lid.

So, take your time. Do your homework. Trust your instincts (and your IT guy). Because once you find the right tool, it won't just enhance your operations, it'll become a quiet, efficient part of your daily rhythm. Like a well-oiled machine. Or a really good playlist.

Overcoming Integration Challenges: From Theory to Practice

(Or: When the Puzzle Pieces Don't Fit and You Consider Setting the Table on Fire)

You've picked your tools, lined up your roadmap, and now it's time to make everything work together. Simple, right?

Well... not exactly.

Integrating AI into a business is a bit like assembling a jigsaw puzzle, only to discover halfway through that some of the pieces belong to another box. Or maybe a different century. Frustrating, but not impossible.

Anticipating the "Oh No" Moments

The first common hiccup? **System compatibility**. New AI tools don't always get along with your existing tech stack, especially if you're working with legacy systems that still think it's 2003. Trying to integrate these can feel like asking a gramophone to connect to Wi-Fi.

Then there's the human element, always the trickiest piece. Change can be scary. People fear replacement, irrelevance, or just plain confusion. If your team starts whispering about HAL from *2001: A Space Odyssey*, you've got some work to do.

Smoothing the Transition

To keep the wheels from falling off, you'll need a strategy for mitigating those oh-so-predictable bumps.

Start with **training**. Offer sessions that actually speak your team's language, not tech jargon, but simple, supportive sessions that explain "Here's what this does and why it matters." Help them move from skeptics to advocates. Show them that AI isn't here to replace them, it's here to make their lives less repetitive and less soul-crushing.

Regular communication is another lifesaver. People fear what they don't understand. So explain, repeat, reframe, and explain again. Offer updates, open forums, and above all, honesty. When people feel informed, they feel in control, and that makes all the difference.

Phone a Friend: Bringing in the Experts

Sometimes you'll need backup. **External consultants** and AI specialists can make integration smoother, faster, and, crucially, less filled with panicked Googling. They've done this before. They know the pitfalls. They bring the metaphorical machete for hacking through the integration jungle.

Outsourcing the trickier stuff frees up your internal team and ensures the job gets done right the first time. It's like hiring a plumber instead of watching ten hours of YouTube videos and hoping for the best. Trust me, your server closet will thank you.

Monitor, Adjust, Repeat

Once everything's in motion, set up **monitoring systems** so you know what's actually happening. Think of it like the dashboard in your car. You want to know before the engine light turns into smoke.

Track performance. Gather data. Don't just trust your gut, back it up with real feedback. Then make adjustments. Improve. Refine. Evolve.

Integration isn't a one-and-done affair. It's a living, breathing part of your business operations now, and like any good system, it needs regular checkups.

Slow and Steady Wins the AI Race

AI implementation shouldn't feel like flipping a giant switch. Roll it out slowly. Incrementally. One department at a time, if needed. This gives teams a chance to adjust and avoids the full-on revolt that comes from too much change too quickly.

Match each AI function with a real business goal. Align your rollouts with strategic needs, not the flashiest features or whatever the tech blogs are buzzing about this week.

Keep talking. Keep listening. Keep learning.

The Real Secret to Smooth AI Integration

Here it is, the unglamorous but golden truth: the real key to overcoming integration challenges isn't found in code or flowcharts, it's found in people.

It's the manager who reassures their team. The tech support person who explains things twice without sighing. The employee who gives honest feedback. The leader who says, "Let's test it first."

Integration succeeds when humans and machines learn to work together. And that takes time, patience, and just a smidge of optimism.

So don't panic if the puzzle pieces don't quite fit at first. With thoughtful planning, clear communication, and a bit of outside help, they eventually will.

And when they do? You won't just have a functioning system; you'll have a better business.

Minimizing Disruption: Aligning AI with Current Processes

(Or: How to Invite AI to the Party Without Letting It Knock Over the Punch Bowl)

Bringing AI into your business is a bit like introducing a new star player to a well-practiced sports team. You don't want them barging in and demanding everyone switch to their playbook; you want them to slip seamlessly into the rhythm, elevate the performance, and maybe win you a few more games without breaking the team spirit.

The best way to do this? **Go slow. Go steady. And for heaven's sake, don't press every button at once.**

Start with small, phased rollouts. Think of it like dipping your toe into a pool before executing a cannonball. Begin in one department, test how things go, make a few adjustments, then move on to the next. It's much better to iron out the kinks in your customer service process than to discover your entire logistics operation has come to a screeching halt because someone misunderstood what "automated" meant.

Marry Your AI to Your Business Goals (Not Just Your Budget)

Here's the thing: AI shouldn't be treated like a shiny new gadget you bought because everyone else was getting one. It's not a side hustle. It's a strategic partner. A co-pilot. The Samwise to your Frodo.

So ask yourself: What are we trying to achieve here? Faster service? Better inventory control? More insightful analytics that don't look like a NASA control panel? Then make sure your AI tools are *actually helping with that*, not just solving problems you don't have.

When your AI initiatives are aligned with your larger business goals, they become part of the momentum, not a detour. It's the difference between strapping a jet engine onto your business and wondering why all your files are suddenly on fire.

Talk Early, Talk Often

Let's talk about communication, or, more accurately, let's keep talking about it. Because nothing derails a smooth AI rollout faster than an office full of people whispering, "Wait, are *we* supposed to be using that now?"

Keep your communication lines open and flowing, like a good dinner party where everyone knows what's being served and nobody's accidentally allergic to the dessert. Regular updates, quick debriefs, team huddles, and perhaps the occasional celebratory donut (or at least a well-timed GIF), these all help keep everyone looped in and onboard.

But don't just talk *at* people. Open up feedback channels too. Let your team share what's working, what's confusing, and what's making them quietly scream into their coffee mugs. You'll uncover useful insights, nip problems in the bud, and, bonus, transform those wary employees into AI cheerleaders faster than you can say "automation protocol."

Step Back and Review the Replay

Once AI is officially up and running (cue the subtle jazz hands), it's time to pause and assess. Think of it like checking box office numbers after a movie premiere. Did it deliver on its promise? Are the reviews glowing or... politely silent?

Check your original performance indicators like productivity, speed, error rates, or customer wait times. But also listen to the mood in the office. Are people adapting? Grumbling less? Logging off earlier with their sanity intact? Good signs.

Gather the data, but don't forget the human feedback. A well-integrated AI system should improve both the **work** and the **workplace**. If you're getting efficiency *and* a happier team? You're not just winning; you're doing a victory lap.

The Big Picture (And Why It Matters)

The real story here isn't about software, dashboards, or tech trends. It's about transformation. AI, when done right, is not just another tool; it's a cultural shift. It changes how your team works, how your processes evolve, and how your business adapts to the future.

But here's the trick: do it without breaking the rhythm. Align AI with existing processes that already work well. Use it to sharpen your edge, not reinvent the wheel. Keep your people informed and involved and build a workplace where innovation doesn't feel like a threat; it feels like progress.

By minimizing disruption, aligning AI with your strategic goals, and keeping communication front and center, you're not just keeping pace with the world; you're quietly setting the pace for others to follow.

In Conclusion (But Not Goodbye)

Let's not kid ourselves, integrating AI takes planning, coordination, and a bit of courage. But it also offers incredible opportunities. With the right roadmap, the right tools, and the right mindset, you're not just updating your tech stack, you're future-proofing your business. Oh, and you're freeing up your time and the time of your entire team. That's a win-win!

And here's the real kicker: it doesn't have to be perfect. It just has to be intentional.

So, whether you're automating simple tasks or streamlining entire departments, remember that AI works best when it's part of your team, not a mysterious machine in the corner humming ominously.

Up Next: Turning Customers Into Fans

In the next chapter, we're going to explore how AI can completely transform the way you interact with your customers. From personalization that actually feels personal (not creepy) to seamless service delivery that makes your brand stand out, we'll dig into how AI isn't just changing business processes, it's rewriting the customer experience.

So, grab a fresh cup of coffee and get ready. This is where things get really interesting.

Chapter Three

AI Applications Across Industries

Why You Should Care About What AIs Doing in Other People's Businesses

NOW, YOU MIGHT BE wondering, "Why on earth should I read about what AI is doing in hospitals or retail stores when I run a landscaping business, or a consultancy, or an online pickleball accessory shop?" Fair question. The answer is simple and a little bit exciting: because innovation doesn't happen in a vacuum, and some of the best breakthroughs are borrowed.

One of the delightful things about AI Agents is how adaptable they are. Like a good Swiss Army knife or a decent pair of walking shoes, they travel well. You'll find them managing inventory in a warehouse one moment, handling customer service in a hotel the next, and then, rather mysteriously, identifying early signs of illness from X-rays while you're still finishing your coffee. They're a bit like that one friend who seems to be good at everything and still somehow has time to compost.

So, even if you're not in healthcare, finance, or retail, you'll want to peek over the fence and see what's happening next door. Why? Because these examples reveal how AI Agents solve real-world problems, and more importantly, how you can adapt those same principles to your own business. It's not about copying someone else's system, it's about seeing what's possible, then customizing the concept to suit your needs.

This chapter is your guided tour through a series of AI-powered success stories, practical applications, and mind-expanding "I didn't know you could do that" moments. Think of it as flipping through a recipe book from a cuisine you don't normally cook. You might not follow every step, but you'll walk away with some clever ideas, a few time-saving tricks, and maybe even the inspiration to try something wildly new.

By the end of this chapter, you'll have a clearer picture of how AI Agents are transforming the way businesses operate, from the back office to customer service and everything in between. More importantly, you'll be equipped with fresh insight into how these same tools can be applied to your entrepreneurial journey, no matter what the industry.

So, let's jump in. Whether you're a dentist, a dog groomer, or a donut mogul, there's something in here for you. Oh, and if you prefer, feel free to skip around.

Revolutionizing Healthcare: AI Applications in Medical Practices

Let's say you walk into a doctor's office bracing for the usual symphony of rustling papers, blinking fluorescent lights, and receptionists negotiating with fax machines like they're hostage negotiators. But instead, you're greeted by seamless, almost eerie efficiency. There's no clipboard. No "Have a seat, we'll be with you shortly." Instead, an AI assistant has already checked you in, flagged your symptoms, and maybe even nudged the doc with a gentle, "Hey, remember to ask about their family history of heart disease."

Welcome to the not-so-distant future of healthcare, where artificial intelligence isn't just a shiny buzzword tossed around at conferences but an actual, behind-the-scenes MVP. From diagnostics to patient care, AI is transforming the medical landscape with the precision of a Swiss watch and the stamina of a grad student on espresso.

AI as Your Diagnostician (Minus the Bedside Manner)

Let's start with diagnostics. Picture AI as the Sherlock Holmes of healthcare, except instead of a pipe and a fiddle, it's armed with machine learning algorithms and a disturbingly accurate ability to read your MRI. These AI-driven imaging tools can detect early signs of diseases like cancer with a level of attention that would make most humans feel like underachievers. No missed lunch breaks. No distractions. Just relentless data digestion and pattern detection.

And then there's predictive analysis, the closest thing we've got to a crystal ball that isn't sold on late-night infomercials. These tools sift through mountains of patient data, spotting patterns and risk factors that even seasoned physicians might miss. It's not magic; it's math. And it's changing the game.

Administrative Relief: AI to the Paperwork Rescue

Now let's wade into the administrative swamp, the part of medicine where doctors spend more time staring at screens than healing humans. Here, AI is like that friend who shows up with pizza and a solution just when you're about to lose your mind.

Appointment scheduling? Automated. Patients can book, reschedule, or cancel without entering the seventh circle of voicemail hell. Claims processing? AI systems can tear through insurance forms faster than a team of clerks on a deadline, dramatically reducing human error and wait times. The result? Doctors get to doctor. Patients get seen. And paper finally stops winning.

Personalized Patient Care: Medicine Meets Custom Tailoring

Next up: personalized care. Think of AI as a master tailor, stitching together treatment plans based on the unique measurements of your DNA, lifestyle, and health history. No more one-size-fits-all prescriptions. These AI systems analyze your genetics, diet, activity level, even how well you sleep, and craft treatment options as bespoke as a Savile Row suit.

Meanwhile, AI-powered remote monitoring tools keep tabs on vital signs 24/7. So instead of waiting for the next checkup, or worse, a health crisis, doctors can intervene early. It's like having a nurse who never naps and a health alert system that never clocks out.

Research and Development: The Unsung Hero of Lab Coats Everywhere

Finally, in the high-stakes, caffeine-fueled world of medical research, AI is changing the pace. It combs through astronomical amounts of data to identify promising compounds for new drugs, drastically shortening timelines that used to take years. AI doesn't need to sleep, snack, or stare into the abyss of Excel. It just works. Constantly. And with growing precision.

In genomic research, AI helps decode the staggering complexity of the human genome, matching people to targeted therapies that fit them better than their favorite jeans. We're talking real personalized medicine here, not the vague kind that promises to "boost your wellness" with kale and turmeric.

So, what does all this mean for you, the business-savvy reader with one eye on innovation and the other on practicality? It means the healthcare industry isn't just dipping its toes into the AI pool. It's doing cannonballs. And whether you're in healthcare or simply inspired by how AI is transforming lives (and bottom lines), there's a lot to learn from the medical world's leap into intelligent systems.

Next stop? We'll look at how AI is stirring things up in a slightly less sterile environment, retail. Spoiler: it involves fewer latex gloves and a lot more algorithms predicting when you'll run out of coffee.

Case Study: AI in Genomic Research

Let's pause for a moment and consider a rather astounding case. In a lab somewhere, a group of researchers, armed with microscopes, determination, and likely a serious caffeine habit, handed their genomic data to an AI system. What followed was less like science fiction and more like science on fast-forward. This digital assistant, let's call it Sherlock Circuits, combed through mountains of data in record time, spotting patterns and correlations that would have taken human researchers' years to detect (and possibly several emotional breakdowns along the way).

The result? A significant leap forward in cancer treatment. Not the kind of vague "better outcomes someday" improvement you see in fundraising brochures, but highly targeted therapies tailored to individual genetic profiles, like bespoke suits, only with a lot more life-saving potential.

And that, in a nutshell, is how AI doesn't just *enhance* healthcare. It reinvents it.

By sharpening diagnostic accuracy, tidying up the cluttered world of administrative tasks, crafting tailor-made treatment plans, and supercharging research and development, AI isn't quietly assisting in the background. It's essentially building a whole new stage for modern medicine. In this bold new era, AI doesn't just stand for *artificial intelligence*; it might as well mean *astonishing innovation*.

As you settle into your next routine check-up, spare a thought for the invisible digital assistant likely humming along in the background. No white coat. No stethoscope. Just pure algorithmic muscle, working 24/7 to help your doctor be even better at their job.

AI in Finance: Automating Complex Analyses

Now let's turn the page to a world powered by decimal points and deadlines: finance. If healthcare is the operating table of AI innovation, then finance is the trading floor, complete with suits, chaos, and that unmistakable scent of ambition.

In this high-speed world, where fortunes are made or lost faster than you can say "Dow Jones," AI is the new power player. Picture yourself as a financial wizard (cape optional), only instead of a crystal ball, you're clutching a neural network capable of spotting market patterns before your morning coffee's even cooled.

AI tools digest entire continents of data in milliseconds, detect trends, and predict shifts with eerie precision. This means decisions once made on gut feelings, or after hastily reading a finance blog at 2 a.m., can now be based on solid, real-time insight. Whether you're evaluating lending risks or deciphering market turbulence, AI is your data-powered copilot.

And fraud? Poor thing doesn't stand a chance. Well, maybe a chance, but AI is making it much more difficult for them.

Real-time transaction monitoring systems powered by AI are like a team of Sherlock Holmeses working around the clock, scanning every blip in the banking matrix for signs of foul play. These anomaly-detection algorithms are constantly learning, evolving, and generally making life miserable for scammers. If someone halfway around the world tries to buy a yacht using your credit card, AI will likely catch them before they've finished typing in the billing zip code.

Meanwhile, over in the high-stakes world of investing, AI has become the portfolio manager of your dreams, relentless, informed, and miraculously free of bias. It assesses your goals, appetite for risk, and market conditions to create a dynamic, diversified portfolio that's about as custom-tailored as it gets. Add machine learning to the mix, and you've got trading algorithms that can respond to the markets in milliseconds. Human brokers, understandably, are both impressed and slightly alarmed.

Even customer service is getting a digital makeover. AI chatbots now field routine banking questions with the cheerfulness of a new hire and the stamina of a machine (because, well, they are one). Need to check your balance at 3 a.m.? Done. Want a mortgage rate comparison while waiting in line at the grocery store? Go for it. Virtual financial advisors are also stepping in, offering personalized financial tips that may not include free coffee, but do provide some surprisingly savvy advice.

This financial metamorphosis isn't about replacing human workers; it's about handing them a better toolkit. AI handles the grunt work, leaving people free to make smarter, more strategic decisions. It's not about machines taking over the economy; it's about giving your business (and your bank account) a serious upgrade.

Up next, we'll explore how AI is transforming one of the most familiar (and finicky) arenas of all: retail. Think smarter shopping carts, eerily accurate product suggestions, and checkout lines that vanish faster than your self-control at a clearance sale.

Quiz Break: AI in Finance - What Goes Where?

Let's test your financial AI savvy. Think of this as a pub quiz, but instead of beer and trivia about 18th-century poets, you're matching AI tools to modern banking dilemmas. Ready?

Match the scenario with the right AI solution:

1. Detects unusual activity in real-time transactions:

2. Personalizes investment strategies based on your risk tolerance:

3. Provides 24/7 customer support for basic inquiries:

4. Predicts market trends using massive data sets:

(Psst... no pressure. If you mess it up, the AI won't judge you. Yet.)

AI's influence in finance isn't about replacing humans with cold, blinking boxes. It's about giving decision-makers access to more insight, less guesswork, and the kind of lightning-fast analysis that makes even the best spreadsheet jockeys weep with envy.

We're talking about predictive analytics that don't just hint at trends, they shout them out. Fraud detection systems that act like financial bodyguards. And investment tools that sculpt your portfolio with the grace of a Michelin-starred chef plating duck confit.

In short, AI isn't just changing how finance is done; it's elevating it. Whether you're a hedge fund titan or a small business owner wondering if you'll make payroll next Friday, these tools give you the ability to plan smarter, act faster, and sleep just a little easier.

So, if you're still waiting to hop aboard the AI express, now's the time. With each algorithm and interface, you're not just streamlining operations, you're future proofing your business in a world where adaptability is currency, and hesitation is costly.

AI in Retail: Shopping Gets Smart

Now, let's waltz into the world of retail, where shelves are stocked, customer tastes are mercurial, and nothing can ruin a perfectly good day faster than a surprise inventory shortage. AI, as it turns out, is the superhero you didn't know your store needed.

Imagine strolling into your favorite shop only to be greeted by an AI-powered system that already knows your favorite shade of blue, the fact that you never buy anything without a discount, and your tendency to impulse-buy socks. It's like being gently stalked by someone with impeccable taste.

These recommendation engines are the digital equivalent of your most fashion-forward friend, the one who always knows what goes with what and never tires of helping you shop. They scan browsing patterns, past purchases, and even social media likes to create a shopping experience that feels, dare we say it, fun again.

But AI isn't just playing stylist. It's also tidying up behind the scenes like a neurotic, but extremely helpful, butler. Inventory management systems, powered by AI, can predict what stock to order, when to order it, and in what quantity. No more guessing whether you need 200 blue sweaters or just 12. These systems read market trends the way your aunt reads horoscopes, frequently, and with uncanny accuracy.

Think of predictive analytics as the retail equivalent of a weather forecast, only instead of rain, it warns you when demand is about to storm in. And automated stock tracking? That's your tireless librarian, quietly keeping tabs on every SKU while you focus on keeping customers happy.

Even the in-store experience gets an AI glow-up. Smart mirrors now let shoppers "try on" clothes without needing to strip down in a poorly lit cubicle. AI kiosks answer questions faster than you can say "Do you work here?" And checkout systems use facial recognition or mobile scanning so slick, you'll wonder if you just walked into a sci-fi film set.

Behind the curtains, AI conducts the supply chain like a maestro. Machine learning tools analyze delivery routes, vendor reliability, and seasonal patterns to keep your operations humming along like a well-tuned orchestra. Problems are spotted before they snowball, and inefficiencies are quietly shown the door.

AI in retail isn't just about bells and whistles. It's about creating experiences that are seamless, personal, and, dare we say, even enjoyable. From the backroom to the fitting room, AI makes everything work better, smarter, and faster.

And the best part? You don't need to be Amazon to take advantage of it.

Whether you're a boutique owner with a nose for quality or a big-box retailer managing an inventory the size of a small town, AI offers tools that put you squarely in the driver's seat. Data becomes insight. Insight becomes action. And action, in this case, means better sales, happier customers, and fewer logistical headaches.

In our next section, we'll explore AI's role in education, where algorithms are stepping in to help students learn faster, teachers work smarter, and the chalkboard finally retires with dignity. Stay tuned.

AI in Manufacturing: Streamlining Production Processes

Would it be confusing if you were to walk into a factory that hums like a Swiss watch, only louder, sleeker, and somehow entirely devoid of grease-stained clipboards or anyone yelling about a broken part, instead of chaos and clangs, you find an orderly ballet of machines moving with purpose, would you find that common-place? It's all rather mesmerizing, like watching a high stakes cooking show where every robot knows exactly when to saute, flip, or plate, and not a single thing gets burned. Welcome to the AI-powered future of manufacturing.

At the heart of this mechanical marvel is a technology known as Robotic Process Automation, or RPA. Think of it as the world's most focused employee, one that doesn't need lunch, never calls in sick, and couldn't care less about office gossip. These robotic systems handle repetitive tasks, like assembling tiny parts, with a kind of obsessive precision that would make a Swiss watchmaker blush. And the best part? They do it all without a single coffee break.

But AI's contributions aren't just about swapping out human hands for robotic ones. It's also about predictive maintenance, which is a fancy way of saying "fixing things before they break." Picture this: you've got machines that talk to you, not in a creepy sci-fi way, but in helpful little nudges. "Hey, just so you know," they might say (if they could), "that motor's going to quit in about 27 hours." This is thanks to algorithms that sift through mountains of sensor data to catch subtle changes that even the best maintenance crew might miss. It's like having a sixth sense for machine trouble, only it's powered by math, not mysticism.

Now let's talk about quality control. Traditionally, this job went to a sharp-eyed person with a clipboard and a weary look. But now? AI systems equipped with machine vision perform tasks faster, more accurately, and without ever being distracted by texts from their spouse or significant other. These systems scan products like laser-eyed hawks (or perhaps very tidy eagles) and flag defects before they make it anywhere near a shipping box. That means fewer returns, fewer customer complaints, and fewer late-night calls about "why does this widget look like it melted in the sun?"

Behind the scenes, AI is also working its magic on the supply chain. If you've ever played the game of "Do we have enough parts to make this thing, or too many parts and nowhere to put them?" then you know what I mean. AI analyzes market trends, order histories, weather patterns, and possibly even the mood of the stock market to figure out exactly how much inventory you need and when you need it. No more guessing. No more warehouses stuffed with the wrong parts. Just smooth, graceful logistics like a ballet choreographed by a very nerdy, very clever robot.

And lest we forget the humans, AI is doing its part there too. On the safety front, AI systems monitor factory floors like lifeguards at a busy pool, alerting staff to potential hazards before they turn into real ones. These systems don't blink, get tired, or sneak off for a smoke; they're always watching, and that's a good thing when forklifts are zooming past at suspicious angles.

Then there's training. In the old days, this meant shadowing someone and hoping not to press the wrong button. Now, thanks to AI-powered virtual reality, workers can practice on virtual equipment until they know it inside and out, all without risking an actual explosion. It's immersive, safe, and honestly, kind of fun.

The takeaway here is that AI isn't just changing manufacturing; it's reimagining it from the factory floor up. It's making everything faster, smarter, safer, and more adaptable. Whether you run a boutique workshop or a multinational manufacturing empire, these tools are opening up new ways to reduce costs, improve quality, and keep your business nimble in an ever-changing world.

So yes, the machines are rising, but instead of plotting our downfall, they're helping us build better widgets, on time, and under budget. And that, I think we can all agree, is a much happier ending.

AI Solutions for the Service Industry: Efficiency and Innovation

What if you're at a hotel, it's late, and you've somehow managed to forget your toothbrush and request extra towels. Instead of summoning the courage to call the front desk and explain your bedtime dental predicament to a stranger, you simply type a message into a little AI-powered concierge app. Seconds later, everything you need is in route, no awkward conversations required. It's hospitality without hassle, and it's just one example of how artificial intelligence is sneakily revolutionizing the service industry while we sleep.

AI, in this sector, behaves somewhat like the world's most eager intern: always available, surprisingly competent, and refreshingly immune to exhaustion. Chatbots, those tireless little conversationalists, now handle everything from booking reservations to fielding complaints about the minibar being too mini. They work 24/7, respond faster than you can say "I need help," and never once say, "Let me transfer you."

But the charm of AI doesn't end with robotic politeness. These systems are also becoming connoisseurs of customer sentiment. Thanks to AI-driven sentiment analysis, businesses can now sift through reviews, tweets, and feedback forms with all the finesse of a Victorian tea leaf reader, except backed by data and devoid of guesswork. It's as if your business now has emotional sonar, detecting delight and disappointment in real time, and gently nudging you to fix things before anyone starts throwing digital tomatoes.

Now, let's talk about personalization. AI in the service world is like that one barista who not only remembers your name but also your favorite drink, your dog's name, and which podcast you're listening to. Whether it's recommending a spa package for the perennially tense traveler or suggesting a wine tasting to someone whose social media posts include 42 photos of Merlot, AI's knack for reading the room, digitally speaking, is uncanny. It's not so much customer service as it is customer delight.

Even loyalty programs have received an AI makeover. No longer are they those dusty punch cards that disappear into the depths of your wallet. AI now customizes rewards based on actual preferences and behaviors, creating offers that people *want* to use. It's like getting a coupon for your favorite thing before you even realize you were craving it. Spooky? Maybe. Useful? Absolutely.

Behind the scenes, AI gets even more impressive. Picture an invisible operations manager armed with the scheduling instincts of a Swiss railway conductor. By analyzing patterns of demand and supply, AI ensures that staff are always deployed efficiently, with no one standing idle and no one overwhelmed. It's the kind of behind-the-scenes magic that makes everything appear effortless up front, like a duck gliding across a pond while paddling like mad underneath.

And just when you think it can't get more sci-fi, along come the drones. Yes, actual drones, zipping through the skies delivering meals, toiletries, or forgotten phone chargers to customers faster than traffic could ever allow. It's a kind of futuristic hospitality that feels oddly normal once you've experienced it. And it doesn't stop there. AI-powered virtual service assistants now act as full-blown digital concierges, booking excursions, managing itineraries, and giving you travel tips that don't involve you Googling for two hours.

As we wrap up this chapter, let's take a step back and appreciate what's really happening here. AI in the service industry doesn't aim toward replacing people with machines. Done right, it's going to remove friction in your organization/process. It's about allowing humans to do more of the human stuff, empathy, creativity, connection, while the machines quietly handle the tedious bits with remarkable consistency and, again, zero coffee breaks.

So, next time you get a perfectly tailored recommendation, or your hotel room magically has your favorite snacks already waiting for you, tip your hat (virtually) to the AI working behind the scenes. It's not just changing the game; it's rewriting the rulebook entirely.

Chapter Four

Real-World Success Stories

LET'S BE HONEST: FOR most entrepreneurs, the idea of implementing AI in your business can feel a bit like standing at the edge of a high dive. The water below looks promising, but you're not entirely sure what's under the surface, or if you're about to belly flop in front of everyone.

And so, like any sensible person faced with change, you might ask: *"Is this really going to work for me?"*

That's precisely why this chapter exists.

Because it's one thing to read about how AI *might* improve your business in theory. It's another thing entirely to see how real people, running real businesses, have harnessed it to save money, reduce stress, and grow like a weed in July.

Now, don't worry, these aren't all Fortune 500 behemoths with IT departments the size of Nebraska. Some are startups. Some are service businesses. Some are just a few steps ahead of where you are now. But all of them share one thing in common: they started with a problem - too many customer complaints, sluggish sales, and clunky hiring processes - and they searched for, and found, an AI solution that actually *worked*.

Each case study that follows is a flashlight in the dark corners of your business. You don't need to copy them line-for-line. Just look for patterns, for sparks of ideas. Look for the moment when a founder realized, "Hey, what if a robot did this part for me?"

By the end of this chapter, you'll have a clearer sense of what's possible, and maybe, just maybe, feel ready to climb up that diving board yourself.

Case Study: How a Boutique Online Retailer Used AI to Delight Customers Without Losing Its Soul

Let's drop in on a small-but-mighty online boutique that sells eco-friendly lifestyle products, like organic teas, reusable bags, quirky handmade mugs, and those all-natural soaps that smell like a woodland fairy just gave you a hug.

The owner, Maya, was like many entrepreneurs: visionary, passionate, and completely overwhelmed. Her customer service inbox looked like a crime scene, and no matter how many late nights she pulled or how many part-time helpers she brought in, it was never enough. Questions came in faster than she could blink: *Where's my order? Can I swap the lavender candle for vanilla? Do you ship to Alaska?* And, perhaps most devastatingly: *Hi, I'm still waiting for a reply...*

Maya didn't want to outsource her customer service to a call center that knew nothing about her brand. She wanted her customers to feel like they were still getting the personal touch, but with far fewer 2 a.m. email replies.

That's when she met her new employee of the month: an AI chatbot.

This wasn't some robotic text generator that spouted nonsense or directed people in loops like a bad phone tree. This was an AI agent trained on Maya's FAQ, return policy, tone of voice, and even her favorite emojis. It responded to common questions in seconds, politely, professionally, and with just enough warmth to sound human(ish).

The results were immediate and, frankly, a little startling.

- **Email response time dropped from 36 hours to under 4 minutes.**
- **Customer satisfaction scores jumped nearly 20% in two months.**
- **Maya cut her outsourced support budget by 40%,** without hiring a single new person.

Better still, the AI didn't just answer questions. It started *learning*. It flagged patterns Maya hadn't even noticed, like a spike in shipping issues with one carrier, or a growing interest in her herbal bath kits during full moons (seriously). With that kind of data, she adjusted inventory and messaging, boosting sales without guesswork.

But Maya didn't stop there. She used the time saved to train her team on how to step in when the AI couldn't, personalized thank-you messages, handwritten notes in VIP packages, or surprise discounts sent after a poor review. Suddenly, customer service wasn't a burden; it was a brand-building superpower.

Now, let's be clear: this wasn't some plug-and-play miracle. Maya had to invest time upfront to set it up right. She learned (the hard way) that it's not about tossing a chatbot onto your website and hoping for the best. She mapped out the most frequent questions, updated her policies, and made sure her bot didn't accidentally offer someone 100% off for returning a used soap bar.

But here's the thing: Maya didn't lose her personal touch. She **protected** it.By letting AI handle the repetitive and the routine, she made room for more creativity, more thoughtfulness, and, perhaps most importantly, more sleep.

Takeaway for Entrepreneurs

If you're drowning in customer messages, struggling to scale your support team, or just want to stop being the only one answering emails on Christmas morning, this kind of AI agent could be your secret weapon.

It's not about replacing you. It's about cloning the best parts of your brain and giving it a keyboard.

Now imagine what *you* could do with that kind of breathing room.

Case Study: Predicting What Customers Want Before They Even Ask

Let me introduce you to Jamie, the founder of a mid-sized SaaS company that creates workflow tools for creative agencies. Jamie is one of those rare folks who can juggle Zoom meetings, cappuccinos, and budget spreadsheets all before lunch, and still forget where she left her phone.

For years, Jamie's business grew steadily. Not explosively, but respectably. Sales went up, then down, then up again, like a polite roller coaster that never quite committed to being thrilling. Her marketing team worked hard, her salespeople gave it their all, and yet... something was missing. They were swimming in data but had no idea what to do with it. Leads were slipping through the cracks, and campaigns often felt like well-crafted guesses.

Then one of her team members, a data nerd with a passion for optimization and really strong opinions about burritos, said, *"What if we let AI tell us who to talk to, when, and how?"*

Enter stage left: **AI-powered predictive analytics.**Or, as Jamie later put it, "My business's version of X-ray vision."

They didn't hire a massive consultancy or buy a million-dollar AI doohickey. They simply integrated a predictive analytics tool with their CRM; one designed specifically for small and mid-sized businesses. The setup didn't even require a PhD or a closet full of servers. It just needed clean data, a few smart questions, and a willingness to trust the math.

And the results? Well, let's just say Jamie upgraded from polite roller coaster to express elevator.

- **Lead conversion improved by 27% in three months.**
- **Sales reps were notified *before* a lead ghosted, based on early warning signs in behavior.**
- **Marketing campaigns became pinpoint accurate**, targeting the right people at the right moment with the right message; almost as if Jamie had found a marketing crystal ball... minus the fog.

The real magic was in **predictive behavior scoring**. The AI analyzed lead activity, like email opens, page visits, webinar attendance, and coffee breaks (okay, not that last one), and flagged the warmest leads automatically. No more "spray and pray" tactics. Every sales call was informed, timely, and relevant. Reps had context and confidence, and that meant more closed deals.

And here's the part Jamie loved most: her team finally *had time to think*.Rather than drowning in spreadsheets and chasing cold leads, they focused on strategy, creativity, and long-term relationships. Sales meetings became less about scrambling and more about scaling.

Of course, it wasn't all sunshine and unicorns. They had to clean up some messy data. And there was a moment when the AI recommended targeting a segment of leads that no human would've picked, and it worked. That required a bit of trust (and a stiff drink). But over time, the system proved itself. And soon, no one wanted to go back to the "gut feeling" days.

Entrepreneur Insight

If your sales strategy feels a bit like fishing in the dark, this kind of AI might just be your flashlight, your fish finder, and your tackle box. It doesn't replace your team. It sharpens them. It helps them know where to focus, when to follow up, and how to close with confidence.

Whether you're running a boutique SaaS company, an agency, or even a quirky e-commerce shop, predictive analytics can help you see opportunities before they knock, and in many cases, before they even buy a doormat.

Case Study: How AI Turned a Tangled Supply Chain Into a Well-Oiled Wonder

Let's talk about logistics. Now, I know that might not sound like the most exhilarating topic, somewhere between filing your taxes and assembling IKEA furniture, but stick with me here. Because what's happening in the world of logistics thanks to AI is nothing short of astonishing.

Picture a global shipping company the size of a small country. We're talking about massive cargo vessels, warehouses that look like aircraft hangars, and enough moving parts to make a watchmaker weep. For years, this company wrestled with the usual suspects: delayed deliveries, rising fuel costs, supply chain snarls, and the occasional rogue weather system. In other words, business as usual.

Then came AI.

Not the kind that writes poetry or argues about philosophy, but the kind that quietly, brilliantly streamlines chaos. Now think about replacing your dusty old roadmap with a smart GPS that also knows about road closures, traffic jams, fuel prices, and incoming thunderstorms. That's what this company did, except their "roads" were ocean shipping lanes, and their GPS was a constellation of machine learning models tracking everything from port congestion to geopolitical tensions.

By integrating AI into their logistics systems, they didn't just shave a few minutes off delivery times; they slashed costs, predicted problems before they happened, and optimized routes so precisely it felt like cheating. Shipping delays that used to send whole departments into panic now barely raise an eyebrow. Their vessels burned less fuel, deliveries arrived ahead of schedule, and customers began doing something rare in logistics: smiling.

- **Real-time tracking** turned "Where's my shipment?" from an existential crisis into a simple dashboard update.
- **Route optimization** saved thousands in fuel costs and maintenance.
- **Predictive risk analysis** meant they knew about the storm *before* the weather apps did.

And here's the real kicker: it wasn't just the tech that made it work. It was **who they partnered with**. They didn't try to build everything from scratch or duct tape random tools together. Instead, they worked closely with AI vendors who understood their unique needs, customizing solutions that fit like a glove rather than a straitjacket. Logistics platforms, supply chain analysts, and in-house teams all worked in unison, like a pit crew fine-tuning a Formula 1 car at full speed.

That collaboration is a crucial piece of the puzzle.

AI, as powerful as it is, doesn't work in a vacuum. You can't just download an app and expect miracles. You need strategy. You need the right people at the table. And perhaps most importantly, you need to know what problem you're actually trying to solve.

Entrepreneur Insight

Whether you're managing a warehouse full of dog treats or trying to get artisan candles delivered before the holidays, **logistics is where fortunes are made... or lost**. AI won't magically teleport your products across the globe (yet), but it will help you move faster, smarter, and cheaper.

The lesson? You don't need to be a global powerhouse to benefit from these tools. Even small businesses can plug into AI-powered logistics platforms to predict inventory needs, track shipments, and keep customers happy without losing sleep (or stock).

As with most good things in business, it's not about doing more. It's about doing smarter.

Case Study: How a Marketing Agency Turned AI Personalization Into Pure Gold

Let me take you inside a modern marketing agency, not the kind with stuffy conference rooms and generic coffee, but one brimming with energy, wall-to-wall deadlines, and the faint scent of caffeine-fueled ambition (in case you haven't noticed, I like coffee). They were good, great, even, but they had a problem that's all too familiar: their campaigns, despite their best efforts, were still landing like polite handshakes instead of bear hugs.

Then they met AI.

Now, this wasn't the kind of AI that writes haikus or debates Nietzsche. This was the laser-focused, data-devouring, customer-segmenting variety. The agency started with **AI-powered customer segmentation**, and it was as if someone handed them X-ray vision. Suddenly, they weren't just tossing out ads and hoping for clicks, they were speaking directly to the right people, with the right message, at exactly the right time. It wasn't just marketing, it was matchmaking with metrics.

Imagine sending a campaign that doesn't feel like marketing at all but more like a well-timed, thoughtful suggestion from a friend who just gets you.

And just when you thought it couldn't get any slicker, AI added another trick: **real-time campaign adjustments**. The software didn't just launch a campaign and walk away. No, it babysat every click, watched every scroll, and made changes mid-flight if something wasn't working. It was like having a stage manager who could rewrite the script during opening night, without anyone noticing.

The agency began personalizing content with a level of finesse that felt borderline psychic. AI didn't just know customers' names, it understood their interests, behaviors, moods, and even what they *might* want next week. Think of it like trading your marketing megaphone for a finely tuned violin that played each customer's favorite tune. Engagement shot up like a rocket on launch day. Customers weren't just interacting, they were responding, sharing, and, best of all, **buying**.

- **Conversion rates? Through the roof.**
- **Customer loyalty? Off the charts.**
- **Return on investment? Enough to make the finance department start high-fiving.**

But the real story here isn't just about better metrics, it's about smarter strategy. The agency used AI not to replace their creative brains but to **amplify them**. They didn't stop at what worked last month. With predictive tools, they started generating content *before* customers even searched for it. They weren't chasing trends anymore; they were creating them.

And here's the kicker: they weren't some billion-dollar conglomerate with an army of engineers. They were a nimble, adaptable agency that dared to ask, *What if we let the data lead the way, and used our creativity to bring it to life?*

Entrepreneur Takeaway

Whether you're running a solo consulting firm, a scrappy startup, or an established business looking for new edge, **AI personalization can be your secret sauce**. It's not about losing your voice, it's about **fine-tuning it so that it resonates every time**.

Start small. Use AI to segment your audience more meaningfully. Then let it help you test and adjust campaigns on the fly. As you gain confidence, explore content personalization that feels more like service than sales. And yes, there's gold in the ROI, but the real reward is in knowing your message is landing where it matters most.

The future of marketing isn't just automated. It's intelligent, adaptive, and yes, a little bit magical.

Case Study: How One HR Department Hired Faster, Smarter, and Kinder... with AI

Let's set the scene: a large corporation, humming with HR energy. Think bustling airport terminal meets caffeine-fueled paperwork frenzy. Phones ringing. Résumés stacking. Recruiters squinting at cover letters like they're decoding hieroglyphics. It was organized chaos, until AI showed up and quietly turned everything on its head.

No fanfare. No trumpet blast. Just a humble algorithm that sifted through resumes faster than you could say "qualified candidate." The company rolled out **AI-powered resume screening and candidate matching**, and the results were startlingly efficient. This wasn't just fast, it was unnervingly precise. The system zeroed in on top-tier candidates with the skill of a job-market Cupid, matching qualifications to job roles like a dating app designed by a Swiss engineer.

And then came the **interview scheduling,** once a logistical nightmare that rivaled holiday travel plans. Suddenly, with AI on the case, interview slots got booked in seconds, not days. No more email tennis. Just clean, coordinated scheduling bliss. Even **post-interview feedback** was automated, meaning HR could finally focus on big-picture thinking instead of drowning in calendar invites and note-taking.

The result? **Recruitment efficiency shot through the roof.** Time-to-hire metrics fell like a rock, and staffing costs did a cheerful little nosedive. It was like discovering a magic espresso that made the whole department faster, sharper, and less irritable.

But the story doesn't end with hiring. Because, as every weary HR veteran will tell you, **recruitment is only half the battle**. Retention, the art of keeping good people, is where the real game is played. Enter AI again, now wearing a different hat.

The company used **AI-driven engagement tools** to monitor employee satisfaction in real-time. Not in a creepy, big-brother kind of way, but more like using a digital stethoscope to listen to the heartbeat of company culture. These insights allowed HR to **design personalized career paths**, tailored to each employee like a bespoke suit. The result? Employees felt seen, heard, and, most importantly, valued.

Now, this might be the part where we all get a little nervous. What about bias? What about fairness? What if the AI has a thing against people who prefer Helvetica over Times New Roman?

The company took these concerns seriously. They conducted rigorous **bias audits** to make sure the AI wasn't quietly making decisions based on the wrong signals. They built in **transparency**, so everyone understood how the system worked and why certain decisions were made. No secrets. No black boxes. Just open communication and an open-door policy, even for algorithms.

In the end, what started as a way to speed up hiring became a full-blown transformation of the HR department. **Faster recruiting. Smarter retention strategies. Lower costs. Happier people.** It's hard to argue with that kind of progress.

Entrepreneur Takeaway

If you're still doing hiring and HR the way you did five years ago, you're probably spending too much time, too much money, and way too much patience. AI can help, not by replacing your team, but by **giving them superpowers**.

Start small. Use AI to screen candidates more efficiently. Let it schedule interviews. Add pulse surveys to keep tabs on your team's morale. Over time, you'll find yourself spending less time chasing resumes and more time nurturing talent.

Because in the end, great businesses aren't just built by hiring the right people. They're built by keeping them inspired, engaged, and growing. And AI, when used thoughtfully, might just be your best new hire.

Let's Wrap This Chapter Up

If you made it through all those stories, give yourself a pat on the back, or at least pour yourself a fresh cup of coffee. You've just taken a whirlwind tour of how AI is quietly revolutionizing everything from customer service and sales to HR and logistics. And here's the kicker: none of these companies replaced their people. They empowered them.

That's the magic of AI when done right. It doesn't take the soul out of your business. It removes the sludge. The repetition. The bottlenecks. The *"Why am I doing this at 10 p.m.?"* tasks that slowly drain the joy out of running your dream.

What these real-world examples show is that AI isn't only for the big dogs. You don't need a PhD, a tech team, or a grant from NASA. You need curiosity, a pain point that you're tired of tolerating, and the willingness to experiment.

So, as you digest what you've read, ask yourself: *Where in my business am I still pushing a boulder uphill? And what if AI could help carry the load?*

Onward!

Chapter Five

Avoiding the AI Backfire

How to Use Smart Tech Without Losing Trust

A FRIEND OF MINE, let's call her Susan (because, well, that's her name), owns a boutique marketing agency. She's savvy, strategic, and the sort of person who color-codes her calendar and actually sticks to it. To save time, Susan decided to automate her hiring process using an AI résumé screener. The promise? Faster, smarter, and unbiased hiring. What she got instead was a digital gatekeeper that somehow filtered out every single qualified candidate with actual marketing experience and prioritized people who once ran a lemonade stand in 2003.

It wasn't just embarrassing, it was confusing. And worse, it was expensive. She lost two great potential hires, spent weeks untangling the mess, and had to personally call candidates to apologize. One applicant actually said, "I figured your company didn't think I was creative enough... based on my name." Yikes.

That's when it hit her, and me: **even the smartest tech can backfire if you're not steering the wheel.**

Trust is earned in decades, lost in seconds... especially when AI's involved.

The High Cost of Getting It Wrong

As entrepreneurs, we're constantly on the hunt for leverage, ways to do more with less, to get hours back in our day, or to finally, finally sleep through the night without the 3 a.m. gremlin whispering, "You forgot to email Janet." And AI, with all its promises, seems like the magic wand. Only, sometimes it turns your to-do list into a to-don't disaster.

This chapter isn't here to scare you off AI. It's here to help you use it **without tanking your brand, your client relationships, or your own peace of mind.** Think of it as your friendly field guide to the dos, don'ts, and "dear-God-please-don't-ever-do-that-again" moments of using AI in your business.

So, buckle in. We're going to navigate how to keep your business efficient without becoming the cautionary tale in someone else's keynote speech.

Why Trust Is Your Business's Real Currency

Let's get one thing out of the way: you are not in the business of AI. You are in the business of **trust,** whether you're selling custom hammocks, cybersecurity packages, or coaching programs that help stressed-out founders get their lives back.

Trust is the currency your customers trade in long before they hand over a credit card. And when AI enters the picture, it either strengthens that bond or quietly saws away at it like a cartoon villain with a hacksaw.

Here's a nightmare for you to consider: A customer messages your automated chatbot, trying to reschedule a service. The bot misunderstands and cancels their account instead. Or worse, it responds with, "Congratulations! Your account has been deleted." That's not just a tech hiccup, that's a reputation risk. And unlike a typo or a missed meeting, it's not always easy to explain away. You can't blame a robot, after all, without also blaming the person who installed it.

What's in it for you? Trusted businesses bounce back faster. They retain more clients. They sleep better.

Here's the rub: trust isn't about perfection. It's about predictability. People want to know that your tech and your team will behave consistently, ethically, and like they were raised with decent manners.

AI offers speed and scale, but without human oversight and transparency, it can quietly erode the very thing that made your business thrive in the first place: **your ability to connect.**

So, before you hand over the keys to your virtual assistant or AI-driven lead screener, ask yourself: *Is this going to build trust... or burn it?*

When in doubt, don't ask, "Can AI do this?" Ask, "Will my customer feel good about how AI did this?"

Coming up next, we'll look at a few cringeworthy scenarios of smart tech gone rogue, and how to sidestep them with your sanity (and customer base) intact.

The Sneaky Ways AI Can Betray You (If You're Not Careful)

If trust is the currency of business, AI is that enthusiastic intern who *means* well but occasionally forwards internal emails to clients by accident.

The problem isn't that AI is evil, it's not sentient, it doesn't have a grudge, and it isn't plotting to ruin your quarterly numbers. But it *can* mess things up in very human ways, just without the human instinct to say, "Oops, my bad."

Let's look at a few common traps smart business owners fall into when trying to use smarter tech:

Trap #1: The "Set It and Forget It" Fantasy

You plug in your new AI email responder, feed it a few sample replies, and think, "Done! I've just saved myself 10 hours a week!" Fast forward to three weeks later, and a loyal client receives a birthday email that begins, "Dear [Customer First Name]."

Bold Takeaway: AI requires supervision, especially initially. Automation is only as smart as its last update.

Even the best tools need tuning. Your business isn't static, nor are your clients. What felt warm and personal three months ago might now sound robotic and cold. AI doesn't evolve on its own; it learns what you teach it. If you don't check in, it starts veering off like a grocery cart with one wobbly wheel.

Trap #2: Data Without Boundaries

Let's say your customer service system is now logging everything. Every message, click, and emoji. All in the name of "insight."

Then one day a client asks how their data is being used, and your answer is... a shrug and a prayer.

If you don't know what your AI is doing with data, your customers will assume the worst.

This isn't just about ethics, it's about perception. Entrepreneurs live or die by how trustworthy they appear. And nothing cracks trust faster than a sense of being watched without consent. Be upfront. Be respectful. And for the love of good coffee, don't let your AI hoard data like a dragon sitting on a pile of GDPR violations.

Trap #3: Delegating the Wrong Tasks

Here's where things get sneaky. Just because an AI *can* do something doesn't mean it *should*. Like responding to sensitive customer complaints and or giving investment advice when it was trained on Reddit.

One business owner I know installed an AI live chat tool. It was supposed to handle basic customer inquiries, but one day it confidently told a confused customer that the company no longer offered refunds. That wasn't true, but it sounded *authoritative*. And the client? Furious. The fallout cost them a 5-star review, two clients, and a weekend of damage control.

What's in it for you? Use AI for the heavy lifting, not the tightrope walk. Let humans handle nuance, tone, and anything that requires empathy.

If you wouldn't let a brand-new intern say it on behalf of your company, maybe your AI shouldn't either.

Emotional Check-In: Why This Matters

When you're burned out, busy, or just plain tired of juggling a dozen plates in the air, it's tempting to let tech take over and hope for the best. But tech, especially AI, is not a substitute for leadership. It's a tool. And like any tool, it can help you build trust or accidentally punch a hole in your credibility.

This chapter is about staying human while scaling smart.

Building AI That Doesn't Bite (Your Brand or Your Customers)

Let's face it: the dream of AI is alluring. It's efficient, tireless, and never complains about not getting a lunch break. But much like a well-meaning but overzealous Labrador, if it's not trained properly, it can cause chaos while thinking it's helping.

The key isn't just using AI; it's building AI that behaves. That means teaching it the rules of your business, your brand, and basic social decency. Let's walk through how to do that without needing a PhD in data science or a fire extinguisher for PR disasters.

Step 1: Train It Like It's New on the Job (Because It Is)

You wouldn't hire a new employee, point vaguely at a desk, and say, "Figure it out." (Well, maybe you would, but only once.) The same goes for AI. Before it can make decisions for you, it needs clear examples, boundaries, and feedback.

Bold Takeaway: Your AI is only as good as the information you feed it. Garbage in, awkward silence or angry customer out.

Train it with real conversations, FAQs, tone guidelines, even examples of what *not* to say. Give it personality, but not so much it starts cracking jokes in condolence emails. You want helpful, not horrifying.

Step 2: Give It Guardrails, Not Just Wings

Freedom is great, until your AI decides to rephrase your refund policy as "all sales final, sucker."

Guardrails are the policies, scripts, and override mechanisms that keep your AI from going rogue. This means setting clear do's and don'ts, limits on what it can say, and, very importantly, a big red STOP button (figurative or otherwise) that lets you or a human step in when things go sideways.

What's in it for you? Guardrails protect your reputation, your customer relationships, and your sanity. They're like spellcheck for ethics.

If something feels risky, it probably is. Keep your AI in the safe zone.

Step 3: Make Feedback Loops a Built-In Feature

AI doesn't "know better" until you tell it. Which means if it makes a mistake, and you don't fix it, it will proudly make that mistake forever, like a dog that thinks "sit" means "spin in a circle, then bark."

Regularly review its outputs. Ask your team and even your customers for feedback. Did the chatbot get snarky? Did the scheduling tool triple-book someone again? Don't just fix the problem, teach the AI what went wrong.

AI learns faster when you correct it early and often. Waiting only makes the weirdness permanent.

Think of yourself as the coach on the sidelines. The AI's playing the game, but you're calling the plays.

Step 4: Keep the Human Touch Where It Matters Most

Yes, AI can write emails, answer questions, and even tell jokes (usually badly). But there are moments when only a human can deliver the right nuance: customer crises, onboarding sensitive clients, or anything involving judgment, context, or emotion.

Don't hand over those key interactions to your AI. Use it to support your team, not replace the beating heart of your business.

AI is your assistant, not your ambassador. Let people be the face of the brand, especially when the stakes are high.

Entrepreneur Check-In: Are You Still in Control?

It's easy to let AI creep further and further into your business, especially when it starts shaving hours off your day. But don't lose track of what it's doing, or worse, who it's becoming on your behalf.

Are your tools reflecting your values? Your voice? Your brand's promises?

If not, don't blame the tech. Retrain it. Refocus it. Rein it in like a friendly dog with a habit of knocking over toddlers because smart tech is powerful. but only when it's working *for* you, not freelancing with your reputation.

Callout: "Clarity beats cleverness. Every time."

Keeping Your AI Compliant (Without Losing Your Mind)

Most entrepreneurs would rather eat glass than read a privacy policy. Seriously, there are fewer footnotes in a tax return. But if you're using AI in your business, especially anything customer-facing, you can't afford to treat compliance like an optional spa treatment.

So, here's the espresso shot version of what matters. Sip it slowly or chug, it's all digestible.

Data Minimization: Because Hoarding Is Not a Strategy

Collecting data "just in case" is a bit like stockpiling canned asparagus; you'll regret it later. The principle of *data minimization*: you only gather what you actually need, and nothing more.

Bold Takeaway: Privacy isn't just legal... it's a selling point.

Customers *love* knowing you're not snooping around their personal lives like a nosy neighbor. When your AI only uses what's essential, you build trust without lifting a finger.

What's in it for you? Less data means fewer liabilities, faster systems, and a lower chance of accidentally leaking someone's vacation selfies or medical records. Win-win.

Consent Mechanisms: Making It Easy to Say Yes (or No)

You know those sites where you need a PhD to find the "Reject All Cookies" button? Don't be that guy.

Consent mechanisms should be straightforward. Tell users what you're collecting, why, and what they're agreeing to. Give them an out. And if you're dealing with anything sensitive, like biometrics or location data, get *explicit* consent. No trickery, no tiny print.

Think of consent like a dinner invitation. You wouldn't say, "Show up at 7, and by doing so you agree to wash dishes and babysit my kids."

Privacy by Design: Like Wheels on a Suitcase

The idea behind *Privacy by Design* is laughably simple, so much so that, once you understand it, you'll wonder why it wasn't always this way. It means building privacy protections into your AI systems from the beginning, not bolting them on later like a wobbly afterthought.

It's the equivalent of someone finally putting wheels on a suitcase. Obvious now, wasn't always then.

So, when you're planning new tools, ask yourself:

- Where is data going?
- Who has access to it?
- How long is it kept?
- Can users delete it easily?

You don't need a compliance team the size of a football roster, just a thoughtful checklist and maybe someone who actually reads the terms and conditions.

The Sleep-Better Framework: Don't Be Creepy

Yes, GDPR and CCPA are important. (Europe and California: the joint overlords of digital privacy.) But more than the acronyms, here's the golden rule: **Don't be creepy.**

- Don't track people without telling them.
- Don't record more than you need.
- Don't sell, swap, or "accidentally" expose user data.

If your AI system is doing something that would make a dinner guest raise an eyebrow, maybe don't.

What's in it for you? You sleep better. You avoid fines. And you don't wake up to a trending hashtag you never wanted: SketchyStartup.

So yes, privacy is a headache. But it's also a superpower. When customers know their data is safe with you, they'll come back. They'll recommend you. And they won't need to squint through 37 popups just to buy your thing.

Bias, Inclusion, and Being a Good Human

Let me tell you a quick story about Raj, a founder of a fast-growing SaaS company who built an AI-powered chatbot to streamline customer support. It was smart, polite, and could handle a dozen queries at once. Things were going great until a longtime customer named Denise, who was visually impaired, called in frustrated. She couldn't get the chatbot to work with her screen reader.

At first, Raj thought it was a fluke. But after digging deeper, he realized the AI wasn't built with accessibility in mind. And it wasn't just Denise. Other users, non-native English speakers, older customers, and folks with different dialects were hitting the same invisible wall. The tool wasn't broken. It was just quietly biased.

Now, to Raj's credit, he didn't wave it off. He pulled in advisors, expanded his user testing pool, and retrained the system with more inclusive data. A few months later, not only was Denise back and happy, but so were a whole bunch of new users who hadn't been able to engage before.

And the best part? Customer satisfaction scores went up. So did retention. Turns out being a decent human being is also a pretty solid business strategy.

What Bias Looks Like in Practice

Bias doesn't always show up with neon lights and villain music. Sometimes, it's the quiet stuff: a facial recognition tool that struggles with darker skin tones, a hiring algorithm that favors certain schools, or a chatbot that doesn't understand accents outside the "default" setting.

For small businesses, the danger isn't just bad PR; it's a missed opportunity. If your AI unintentionally excludes people, you're leaving money, trust, and reputation on the table. Sometimes, without even knowing it.

Reflection Prompt: *"Who's not being served by your current system, and how would you know?"*

Inclusive Design = Growth Strategy

Now here's the thing: inclusive design isn't just a checkbox. It's an engine for growth.

When you build tools that work for a wider variety of people, from different backgrounds, languages, ages, and abilities, you're not just being thoughtful. You're expanding your market. You're increasing loyalty. You're creating products that real humans can actually use and love.

If your competitor's tool works for 60% of the population... and yours works for 90%. Who do you think wins?

The Business Case for Diverse Input

When teams include a wider range of voices, from frontline staff to underserved customers, something magical happens. Assumptions get challenged. Edge cases get surfaced. And your AI stops behaving like it was built by a room full of clones.

Diverse perspectives lead to smarter decisions, fewer blind spots, and, yes, better AI. The kind that doesn't just "work," but actually *serves.*

So no, this isn't about being politically correct. It's about being commercially smart *and* human centered. The two aren't at odds; they're powerfully aligned.

Bold takeaway: "If your AI only works well for people like you... it doesn't work."

Tech You Can Trust, Business You Can Sleep With

By now, you've seen the upside and the underbelly of AI. It's dazzling, yes, but also delicate. Like letting a robot babysit your reputation. So, here's the heart of it all:

AI should scale your values, not replace your judgment.

You don't need to be an ethicist or write lengthy treatises on algorithmic fairness. But as entrepreneurs, we *are* the ones who decide how our tools get used. We're the ones responsible for making sure that what we automate doesn't unravel what we've built.

Because at the end of the day, **this isn't just about compliance, it's about trust.** Trust with customers, with your team, with the brand you've worked so hard to grow. And trust, my friend, is the one thing AI can't fake.

So, yes, go ahead and use smart tech. Build automation into your business. But do it in a way that keeps you proud of what you're scaling. Keep your processes fast, but your ethics solid. Be transparent. Be inclusive. And every once in a while, pause to ask, "Would I want to be on the receiving end of this system?"

Because **you can use automation and still be human, and that's your real edge.** Your kindness doesn't have to take a backseat to your KPIs.

Picture this: an AI system that's as trustworthy as your best employee, efficient, sharp, never calls in sick, and totally aligned with your mission. That's not a pipe dream. That's what this chapter just helped you build.

Now go out there and lead with both smart systems and a steady compass. Your business, and your sleep schedule, will thank you.

Final bold takeaway: "Smart businesses don't just automate, they elevate."

Chapter Six

Tackling Time Constraints

Efficient AI Learning for Busy Entrepreneurs

NOW THAT YOU KNOW how to keep your AI from backfiring on your business reputation, let's talk about the real-world speed bumps that trip most entrepreneurs up, like time, budget, and tech headaches.

If you're already juggling a dozen flaming swords labeled "client meetings," "invoicing," and "I forgot to eat again." Then someone tells you to "learn AI," as if it's just another to-do item you can slot between brushing your teeth and world domination.

For most entrepreneurs, the idea of learning AI feels about as feasible as mastering ancient Greek in a weekend, while on fire. You've got dreams of using AI to save time, make smarter decisions, and maybe even get home before dinner, but the path to getting there looks suspiciously like a rabbit hole lined with jargon, dashboards, and... sigh... more logins.

But here's the good news: you don't need a PhD or a 30-hour-a-week study schedule to get the benefits of AI. You just need a smarter way to learn.

What's in it for you? Less time flailing, more time winning. You'll learn enough to make smart decisions, use the right tools, and delegate the rest with confidence.

Bite-Sized Learning for Real Life

Online micro-learning platforms, like Coursera, Udemy, and LinkedIn Learning, offer courses that break AI down into digestible chunks you can absorb during lunch, waiting in line, or hiding in your car for five minutes of peace. These aren't sprawling college lectures; they're more like espresso shots of insight, served up in ten-minute doses.

Think of them as your AI gym: you don't need to bulk up overnight, just get stronger, a few reps at a time.

Bold takeaway: "You don't need to learn everything, just enough to lead well."

Let AI Teach You About AI

Here's where it gets deliciously meta: some AI platforms will actually help you learn AI. Personalized learning engines track your progress and recommend what to focus on next. It's like Netflix, but for your brain, and with fewer true crime documentaries. These tools adapt to your pace and preferences, nudging you along without judgment (or ads).

Want reminders to keep you on track? AI can handle that too. You'll get subtle prompts to study, progress tracking, and even motivation boosts, without feeling like you're being chased by a pushy gym coach.

Protecting Time Like It's Your Wallet

Here's a hard truth: if you don't schedule time to learn, the universe will happily fill your day with other people's emergencies. Carve out protected time on your calendar for learning, just like you would for your most important client. Because, in a way, that's who this is for: your future self.

Frameworks like the Eisenhower Matrix help prioritize what matters most, letting you build learning into your week without burning out. And once you start seeing returns—like understanding which AI tools are worth paying for or delegating smarter—it pays off faster than you'd expect.

Learning With Others is a Superpower

Don't go it alone. Online entrepreneur communities, Slack groups, and AI-focused webinars let you connect with others figuring it out alongside you. You can ask questions without fear of sounding dumb and discover tools that other people have already vetted (and survived). Peer learning isn't just more fun, it's more efficient.

Reflection Prompt: What 30-minute window this week could you protect for learning something that could save you hours next month?

Addressing Cost Concerns: Budget-Friendly AI Solutions

Let's be honest, when most entrepreneurs hear the words "AI implementation," they assume it's going to cost roughly the same as a small yacht. Or at least a new espresso machine they don't really need but suddenly must have. Either way, it sounds expensive. Which is why so many smart, scrappy business owners shove it in the "maybe later" drawer, right next to hiring a full-time CMO or launching a podcast from a beach hammock.

But here's the surprising truth: the AI revolution isn't just for billion-dollar brands with teams of engineers and enough budget to wallpaper their boardroom with hundred-dollar bills. These days, AI tools come in more price points than breakfast cereal, and many of them are free (or close enough to it that your accountant won't raise an eyebrow).

What's in it for you? You get access to powerful tools that can automate tasks, improve decision-making, and save you boatloads of time, without wrecking your cash flow.

Bold takeaway: "You don't need big money to build smart systems."

The Power of Open Source (AKA: The Thrift Store Jackpot)

Open-source AI tools are kind of like finding a designer suit in a thrift shop. They might take a little digging, but when you find one that fits your needs, it feels like a win for both your wallet and your inner tech nerd. Tools like **Hugging Face**, **Scikit-learn**, or **TensorFlow Lite** are free, powerful, and constantly updated by generous developers who apparently run on caffeine and community spirit.

The downside? You might need some help configuring them, but that's still cheaper than shelling out for a full-stack custom solution right out of the gate.

Pay-As-You-Grow: The Cloud-Based Sweet Spot

Cloud-based AI platforms like **Google Cloud AI**, **Microsoft Azure**, and **Amazon's SageMaker** offer scalable pricing, meaning you only pay for what you use. It's like ordering à la carte instead of an expensive prix fixed menu, you're too full to finish. You can start small, automate one process (like scheduling or lead scoring), and scale up as your revenue does.

This turns AI into an operational cost, not a capital investment, which is music to the ears of anyone who's ever sweated over payroll.

Pro Tip: Look for platforms that offer generous free tiers or trial periods. You can test the waters before diving in.

The Magic of Micro-Starts (A.K.A. Don't Boil the Ocean)

You don't have to "AI-ify" your whole business overnight. In fact, please don't. Start with a single pain point: something repetitive, annoying, and low risk. Think: automated email replies, appointment booking, invoice sorting, or simple customer queries.

This small start gives you a win, builds confidence, and keeps costs minimal, like remodeling just the kitchen before gutting the whole house.

Reflection Prompt: What's one task in your business that drives you nuts and could be automated tomorrow?

AI-as-a-Service: Netflix for Automation

If setting up your own AI sounds like more hassle than it's worth, you're in luck. Subscription-based platforms like **Jasper**, **Pictory**, **Zapier AI**, or **Beautiful.ai** offer done-for-you tools for writing content, analyzing data, or creating presentations.

It's like Netflix for automation; you get access to a powerful library of tools for a low monthly fee. And unlike your gym membership, you'll actually use this.

Free Money? Yes, Please.

Governments and nonprofit programs love to hand out incentives for tech innovation. Depending on your location, you may qualify for grants, tax deductions, or startup credits if you're using AI to grow your business or create jobs.

No, it's not always easy to find or apply for. But if the idea of funding your AI journey with someone else's money doesn't get you excited, check your pulse.

Bold takeaway: "Smart AI spending is like hiring a genius intern who never sleeps and doesn't invoice you hourly."

Managing Change – Preparing Teams for AI Integration

Bringing AI into your business is a bit like bringing home a new puppy. There's excitement. There's hope. There's probably someone in your office who's already declared, "This thing is going to destroy everything!" And like puppies, AI needs a little house training, a clear set of rules, and some patient leadership to settle in and thrive.

Because here's the thing: even if the tech is brilliant, if your team's not on board, it won't stick. It'll just be one more tool that gathers digital dust while everyone stubbornly sticks to what they already know.

What's in it for you? Less resistance, more momentum. You get buy-in, smoother adoption, and a culture that actually wants to innovate.

Bold takeaway: "AI won't transform your business, your people will."

Start with the Fear (and Then Calm It Down)

People are, by nature, suspicious of things that beep and make decisions faster than we can. The fear isn't just about job loss, it's about relevance. It's about not wanting to feel dumb or left behind.

So, talk about it. Open the door for honest conversations. Ask: What worries you about this new tool? What would make it feel useful, not threatening? This small act of empathy goes further than any tech demo ever will.

Quick Script You Can Steal:"This isn't about replacing people, it's about removing the parts of your job that are annoying, repetitive, or impossible to keep up with."

Celebrate the First Win (Even if It's Tiny)

AI doesn't need to blow your socks off right away. Maybe it saves the team 20 minutes a day. Maybe it catches a typo. Maybe it automates one email sequence.

Whatever the first success is, shine a light on it. Name it in the Monday meeting. High-five the intern who set it up. These early wins create psychological momentum that sounds like: *Oh, okay... this isn't scary. This is actually kind of nice.*

Bold takeaway: "People support what they help succeed."

Train Like It Matters (Because It Does)

Let's say you show someone a microwave, not telling them how it works, and expect them to cook Thanksgiving dinner. That's what AI rollouts can feel like when training is rushed, skipped, or jammed into a 30-minute lunch-and-learn nobody asked for.

Invest in real training. Not dry manuals or jargon-filled webinars, actual, engaging sessions that show people how AI fits into *their* day-to-day life.

Idea to Steal: Try a "choose-your-own-adventure" approach. Create AI learning tracks by role, marketing, admin, sales, ops, and let team members follow what's relevant to them.

Appoint Your AI Champions

Every team has them: the curious tinkerers, the early adopters, the ones who actually read the product update emails. These are your **AI champions**. Empower them.

Give them early access to tools. Let them test stuff. Ask them to teach others in plain English. When people see peers leading the charge (instead of a consultant in a crisp white shirt talking about "leveraging synergies"), they're more likely to listen.

Reflection Prompt: Who on your team gets excited about new tools, and how can you support them to lead?

Create Feedback Loops (and Actually Use Them)

Once AI is in motion, your job isn't done. You need feedback. Not just metrics, but feelings. Is the tool helping? Where's the friction? What's confusing? What's delightful?

Run a quick monthly pulse survey. Host an "AI wins & fails" coffee break. Use the data to make changes. When your team sees their input shaping the rollout, they stop seeing AI as "theirs" and start seeing it as *ours*.

Bold takeaway: "Feedback isn't a box to check, it's your strategy's secret weapon."

Keeping Your AI Compliant (Without Losing Your Mind)

Most entrepreneurs would rather eat glass than read a privacy policy. And I get it, compliance sounds about as thrilling as filing your taxes while watching paint dry. But here's the thing: if you skip this part, the consequences can be far more painful than a tedious read. We're talking lawsuits, lost customers, sleepless nights, and viral headlines that start with "Local Business Sparks Outrage After..."

So, let's take the espresso-shot version of what actually matters.

What's in it for you? You sleep better. You avoid fines. And most importantly, your customers trust you.

Bold takeaway: *"Privacy isn't just legal, it's a selling point."*

The Rules Are Here... Like It or Not

You've likely heard of GDPR, CCPA, or maybe something more local but just as ominous. These acronyms are shorthand for "You are responsible for what your AI does with people's data."

You don't need a law degree, but you do need to know the basics:

- **GDPR** = Don't use personal data without permission (especially if you're dealing with Europe).

- **CCPA** = Give Californians the ability to say, "No thanks, don't sell my data."

- **Universal Principle** = If it feels creepy, it probably is.

Rule #1: Minimize the Data You Collect

Here's a revolutionary idea: only collect the data you actually need.

You don't need someone's favorite ice cream flavor just to send them an invoice. Don't hoard data like you're prepping for a digital apocalypse. Stick to what's relevant to the service you're providing.

Bold takeaway: *"Less data = less liability."*

Rule #2: Get Clear Consent (Not Hidden Checkbox Consent)

Consent should be obvious, not buried under sixteen paragraphs of legal soup.

Use plain language. Let people know what you're collecting, why you're collecting it, and what you'll do with it. If they say "no," honor it. Don't just hide their info under a metaphorical digital rug and hope nobody asks.

Example: "We use AI to improve customer support. Your chat will be analyzed to help us respond faster. Want to skip that? Click here."

Rule #3: Design Privacy Like You'd Design a Great Suitcase

You know those suitcases with wheels that glide like butter through the airport? They're so obvious now, but it took someone a while to invent them. That's **privacy by design**.

Build systems that *start* with privacy in mind. Don't slap it on later like duct tape over a leaking pipe.

- Limit access.
- Use encryption.
- Let people delete their data.
- Make audit logs your new best friend.

Bold takeaway: *"Good privacy design is like wheels on a suitcase, obvious, elegant, and essential."*

Bonus: Tell People What You're Doing (And Why It's a Good Thing)

Don't just avoid the creepy factor, flip the script. Make data protection part of your brand promise.

- We use AI, but your privacy is never for sale.
- Our AI helps us serve you faster, without snooping into your life.

When you own the conversation, you build trust. And in an era where trust is rare, that's a competitive advantage you can't buy.

Bias, Inclusion, and Being a Good Human

It started with a great idea: a small but growing startup built an AI tool to help businesses sort through job applicants faster. Resume in, score out. The promise? Fair, efficient, data-driven hiring.

Only later did they discover something troubling. Their tool was filtering out women who had taken time off to raise children. Not intentionally, of course, but when they reviewed the AI's "logic," it turns out the system had learned to equate gaps in employment with poor performance, thanks to the data it had been trained on.

That wasn't just a technical glitch. It was a *people* problem. And they knew it.

To their credit, the founders didn't sweep it under the rug. They brought in a diverse review team, re-trained the AI on more representative data, and made the system more transparent. Applicants began receiving fairer reviews, and customers took notice. They weren't just fixing a flaw; they were building trust.

Bold takeaway: *"Bias in, bias out. Inclusive data makes for inclusive AI."*

What Bias Looks Like in Practice

Bias in AI doesn't usually show up wearing a name tag. It's sneaky. It hides in datasets that reflect historical inequalities, or in design decisions made by teams who all look and think alike.

You may notice this when your chatbot misinterprets certain dialects.Or when your product recommendation engine shows completely different suggestions based on someone's name.Or when a facial recognition feature just doesn't recognize every face.

It's not always malicious. But that doesn't mean it's harmless.

What's in it for you? Bias isn't just a PR risk; it's a lost opportunity to serve more people better.

Inclusive Design = Good Business

Designing with inclusion in mind isn't charity. It's strategy.

When your tools work well for more people, more people can use them, and more people become loyal customers. It's not about being politically correct; it's about being *commercially smart.*

- Add captions to videos.

- Make interfaces that are mobile-friendly and readable. \

- Train your AI on diverse voices, languages, and behaviors.

Inclusivity isn't extra work. It's smart work.

Bold takeaway: *"Inclusive design is like curb cuts, made for some, helpful to all."*

Diverse Input = Better Decisions

The more perspectives you bring into your development process, the more blind spots you eliminate. Diverse teams build better products. They ask different questions. They notice things others miss.

Think of it like seasoning a dish; you don't want five people all adding salt. You want balance, depth, and complexity. That's what makes a great meal. And it's what makes a great AI tool too.

When you bring in different lived experiences, backgrounds, and skill sets, you're not just checking a box. You're future-proofing your innovation.

Reflection Prompt:"Who's not being served by your current system, and how would you know?"

Pause here for a moment. Think about the people your AI might overlook. Are you collecting feedback from all types of users? Are you testing for fairness across different groups?

Sometimes the first step to being more inclusive is realizing who isn't in the room, and inviting them in.

When you lead with values, your tech follows suit. And your customers? They feel the difference.

Tech You Can Trust, Business You Can Sleep With

If you've made it this far without breaking into a nervous sweat about your chatbot turning rogue or your AI assistant subtly offending your customers, congratulations. You're already miles ahead of the average entrepreneur who's still trying to figure out if "GPT" stands for some kind of protein shake.

But here's the thing: AI isn't going anywhere. And if you want it to work *for* you—not *against* you—it's time to think about more than just speed and savings. You need to think about trust.

Because here's the not-so-secret secret: AI reflects *you*. Your values. Your intentions. Your blind spots (yes, we all have them). And your judgment.

Bold reminder: *"AI should scale your values, not replace your judgment."*

So, what's the takeaway from all these stories of mishaps, mistakes, and do-overs?

It's not that you need to become an ethics professor or hire a full-time compliance officer who speaks fluent regulatory gobbledygook.

It's that you need to stay human.

Human enough to question where your data's coming from.Human enough to admit when your system's causing harm.Human enough to fix it—and tell people you did.

Because at the end of the day, entrepreneurship isn't just about making something that works. It's about making something that *matters*.

And trust? That's your competitive advantage. It's what keeps customers coming back, employees staying loyal, and your reputation intact even when the tech gets weird.

So go ahead, automate. Delegate to machines. Embrace the robots.

But do it with heart. With intention. With the kind of thoughtful leadership that makes your business not just faster, but better.

Final image to leave you with: Think about your AI setup as the perfect employee. It never calls in sick. It handles tasks efficiently. It remembers names. It makes suggestions that actually help. And, most importantly, it reflects your mission, your values, and your commitment to doing good business well.

That's what this chapter was all about.

Not tech for tech's sake.

But tech you can trust.And a business you can sleep with.

Chapter Seven

Enhancing Business Efficiency with AI

How to Hand Off the Tedious Stuff, So You Can Get Back to the Brilliant Bits

If you're anything like most entrepreneurs, your to-do list doesn't just grow, it breeds. Tasks multiply in the dark like forgotten leftovers in the back of your fridge, quietly transforming into something too unruly to face without goggles and tongs. You wake up with big plans: launch that new campaign, map out next quarter's strategy, maybe even (gasp) take a lunch break. But by 9:14 AM, you're buried under scheduling snafus, customer follow-ups, and a Google Sheet that crashed for the third time this week.

And here's the real kicker: it's not the work that's hard, it's the repetition. The piddly stuff. The "Why am I doing this when I could be doing literally anything else?" stuff.

When Your Day Feels Like a Game of Entrepreneurial Whack-A-Mole

That's where this chapter comes in.

What's In It for You?

This isn't just about automating tasks; it's about buying back your time, your creativity, and maybe even your sanity. We're going to show you how to:

- Let AI handle the boring parts (like data entry and calendar gymnastics)
- Make smarter decisions with real-time insights (no tarot cards required)
- Communicate better (internally and with customers)
- Streamline project management without losing your mind
- And finally, treat your customers like the unique, lovable, complicated humans they are, with personalized service at scale

By the time we're through, you'll have the playbook for transforming your operations from "barely keeping up" to "I've got this." This chapter is your cheat code to running a business that runs itself—at least, the parts that shouldn't need your genius in the first place.

Bold takeaway: *"Efficiency isn't about doing more, it's about doing less of what drains you, and more of what moves the needle."*

Ready to reclaim your brainspace and finally feel like the leader of your business again?

Let's begin.

Automating Routine Tasks

Freeing Up Time for Innovation (and Maybe a Nap)

What if you sit down with a fresh cup of coffee, ready to dive into the kind of big-picture thinking that actually grows your business. Five minutes in, you're sidetracked. A client email needs answering, the spreadsheet won't stop crashing, and Brenda from accounting wants to know, again, if you've uploaded the new onboarding form. (Odds are on that idea that, you haven't.)

Welcome to the entrepreneur's daily rodeo. Only instead of riding wild stallions, you're wrangling repetitive tasks that seem custom-designed to drain your will to live.

Here's the truth: you didn't start a business to become its most overworked employee. And yet, here we are, up to our eyeballs in admin.

Now, let's look at a different scenario. You, relaxed. Your time, reclaimed. Your day, guided by intention rather than interruption. It's not a fantasy, it's what happens when AI rolls up its digital sleeves and starts tackling the busy work.

Meet the Culprits: The Tasks That Steal Your Sanity

We're talking about the usual suspects:

- **Data entry:** Like folding laundry, but somehow more mind-numbing.
- **Scheduling:** A logistical version of Tetris that no one ever wins.
- **Document shuffling:** Searching for a PDF like it's a treasure map to El Dorado.

These are the productivity vampires, draining your time, energy, and enthusiasm in quiet, persistent bites.

AI to the Rescue: Your New Digital Assistant

Now, picture AI not as a robot overlord but as your eager, efficient sidekick. It doesn't complain, it doesn't need lunch, and it never "forgets" to send the email.

Tools like **Robotic Process Automation (RPA)** take on repetitive workflows like invoice entry, approvals, or CRM updates with all the charm of a Swiss watch, reliable, precise, and not prone to Monday moods.

Meanwhile, **AI schedulers** and **calendar tools** coordinate your meetings like the world's most thoughtful executive assistant, one who actually knows your availability and doesn't book a 6 PM strategy call when you're brain-dead and dreaming of tacos.

And if your business relies on forms, reports, or files that need tagging, tracking, or cataloguing, **AI-powered document managers** will do the heavy lifting faster than you can say, "Where's the client agreement from three months ago?"

What's in it for you? Less brain fog. More flow. And, quite possibly, the joy of remembering what your lunch tastes like when eaten sitting down.

Bold takeaway:

"Every task you automate is one more opportunity to think, lead, and create, like you were meant to."

A Moment of Reflection

Grab a pen, a napkin, or your notes app. Ask yourself:

What's the one task I dread doing every week, something I *know* doesn't need my brain?Could that task be automated? Delegated to an AI system? Or at the very least... simplified?

Jot it down. The first step to freeing up your schedule is recognizing where your time is quietly being wasted.

Making Smarter Decisions, Faster

Let's talk about something entrepreneurs are both famous for and terrified by: gut instinct. That sixth sense that says, *"launch it now,"* or *"don't trust that supplier,"* or *"yes, that sock company on TikTok is going to be huge."* Instinct has its place, but let's face it, it's not infallible. And when the stakes are high (which, let's be honest, is most of the time), you want something a bit sturdier than your gut to lean on. Enter AI.

What if your decisions weren't just well-intentioned stabs in the dark but informed, nuanced, and guided by more data than a Wall Street analyst on espresso? That's the magic of AI-driven decision-making. Think of it as putting your gut on steroids, faster, sharper, and backed by cold, hard evidence.

The Magic Bees of Business Intelligence Let's start with the unsung heroes: AI data aggregation tools. These are like industrious little bees, buzzing around your digital ecosystem, gathering nectar from every flower. your website traffic, customer emails, social media buzz, sales receipts, and a dozen other hidden hives of information. The result? A single, digestible honeycomb of insight.

You don't have to be a data scientist or wear a lab coat to benefit. These tools do the legwork, organizing your data so it makes sense at a glance. Patterns emerge. Trends become obvious. Blind spots stop being, well... blind.

Your Crystal Ball (Without the Voodoo) Now let's talk predictive analytics. No, it's not wizardry. It's math. Good math. Predictive AI doesn't just tell you what's happening, it hints at what's likely to happen next. That means you can spot market trends before your competitors even know they exist. You can anticipate customer behavior, adjust strategies, and here's the kicker, do it with confidence.

Visual dashboards bring this all to life, transforming your mountain of numbers into charts and graphs you can actually use. Instead of drowning in data, you're piloting a jet with a clear, easy-to-read cockpit. And yes, there are even dashboards that track your dashboards. That's how meta it's getting.

Upgrading from Gut to Game Plan If you're already using business intelligence (BI) tools, AI doesn't replace them, it supercharges them. It's like your current reports, only smarter, more responsive, and able to spot patterns even your sharpest team members might miss.

AI-integrated BI systems become your consigliere, the business brain that never sleeps. And automated reporting tools? They're like your own analytics butler, delivering custom insights wrapped in a bow every morning.

The Culture Shift: Data Doesn't Have to Be Boring Of course, none of this matters if no one knows how to read the numbers. That's where data literacy comes in. And no, we're not talking about turning your whole team into Excel wizards. It's about giving them the confidence to interpret a trendline, ask better questions, and spot red flags before they become red sirens.

A data-literate team doesn't just react faster, they innovate smarter. Offer training that makes it approachable, less *"spreadsheet bootcamp,"* more *"aha moment."*

Reflection Prompt Think about the last big decision you made in your business. What data did you use to guide it, if any? If your gut was steering the ship, imagine how that decision might have looked with real-time customer insights or market predictions in hand. What would you do differently?

Bold Takeaway"Your gut is great, but AI gives it a map, a compass, and maybe a nice espresso."

Excellent.

Enhancing Communication with AI-Powered Tools - Talk Less. Say More. Sound Brilliant.

If you've ever stared at your inbox like it just insulted your dog, you're not alone. For entrepreneurs, communication isn't just a task; it's a relentless avalanche. Emails, Slack pings, Zoom calls, texts from your team, DMs from customers, and the occasional carrier pigeon from your accountant; your average day feels more like triage than leadership.

But what if communicating didn't feel like a matter of survival? What if, instead of drowning in messages, you surfed them with grace and style, maybe even a little flair? Enter AI.

AI as Your Internal Translator (and Therapist) Let's start inside the business. AI chatbots now handle internal queries faster than you can say, "Where's that file from Q3?" Picture a tireless, never-grumpy assistant who knows the answer to everything, or at least where to find it, and never once says, "That's not my job."

Meeting schedulers, powered by AI, are the unsung heroes of workplace peace. No more back-and-forth emails that feel like negotiating a hostage release. Just clean calendar slots and magically conflict-free meetings.

And for remote teams? AI tools are the duct tape holding the virtual office together. From real-time transcription and auto-summarized meeting notes to task assignments that don't get lost in the chat abyss, it's like having a team coordinator with the memory of an elephant and the efficiency of a Swiss watch.

AI with a Customer-Side Manner Now pivot outward. Your customers are just as overwhelmed as you are. What they want, desperately, is for you to make things simple, personal, and human. Ironically, AI is the tool that helps you do just that.

AI email responders can handle most customer inquiries with tailored answers that sound like you (on a really good day). They learn from past conversations and adjust tone based on customer sentiment. So instead of sounding robotic, they sound like the helpful, emotionally attuned version of yourself, one that didn't just spill coffee on their laptop.

Sentiment analysis tools take this further, detecting when a customer is annoyed, delighted, confused, or about to rage-post (before the renaming/branding of Twitter to X, it would have been "rage-tweets"). This gives you a chance to intervene like a hero, not a fire extinguisher.

Breaking the Language Barrier Without Breaking a Sweat Real-time AI translation is no longer the stuff of spy movies. Whether it's voice, text, or both, these tools now make it possible to serve clients and partners in multiple languages without hiring a dozen interpreters.

Want to impress a client in Barcelona, smooth over a deal in Tokyo, or just understand your international team better? AI makes multilingual fluency a feature, not a hurdle.

And let's not forget accessibility. AI-driven tools that transcribe audio to text, add captions, or convert text to speech make your business more inclusive, legally compliant, and, frankly, a whole lot easier for everyone to understand. Because what's the point of having a message if half your audience can't hear it?

Bold Takeaway"AI won't write your love letters, but it'll make sure they get delivered, translated, and answered before lunch."

Reflection Prompt Think about your current communication pain point. Is it customer response time? Team coordination? Language gaps? Pick one and ask: could an AI tool help here? And if it could—what would you do with the time you just got back?

Streamlining Operations: AI in Project Management - From Herding Cats to Conducting a Symphony

Managing a business project often feels like juggling flaming torches while riding a unicycle, on gravel, blindfolded. Deadlines creep, people forget things, files go missing, and before you know it, you're answering emails at midnight and wondering why your eye keeps twitching.

But what if project management could feel... graceful? Enter AI: your backstage manager, timekeeper, and crisis prevention squad, all rolled into one.

AI as Your Taskmaster (The Good Kind) Project management software powered by AI doesn't just keep a to-do list. It prioritizes tasks based on importance, deadlines, workload, and even resource availability. It's like having a personal productivity coach who doesn't charge by the hour or judge you for forgetting that one crucial deliverable.

Automated timeline adjustments? Pure magic. Instead of manually updating everyone when one piece of the puzzle shifts (again), AI recalibrates timelines automatically. It's like Waze for your workflow, adjusting the route in real time when someone parks a semi-truck across your Gantt chart.

The Great Resource Tetris Solver Allocating resources without AI is a bit like packing a suitcase for a two-week trip in a carry-on. Something always gets left behind.

AI doesn't guess. It calculates. With predictive analytics, you can spot resource shortages before they become bottlenecks. Is your lead designer drowning while three other team members wait around for instructions? AI sees it and flags it before you hear about it during a meltdown.

It's project foresight instead of hindsight, and the difference is peace of mind.

Building Better Teams (Without Group Therapy) Collaboration is where good projects go to die, usually in a thicket of miscommunication, unclear expectations, or someone accidentally deleting the shared folder.

AI-powered collaboration tools help smooth the bumps. Real-time feedback features analyze team interactions (yes, really), surfacing potential friction points or communication breakdowns before they become morale-killing monsters. Think of it like a marriage counselor for your team, minus the awkward silences and trust falls.

Reporting Without Repeating Yourself Most entrepreneurs don't wake up excited about status reports. But AI-generated dashboards might just change your mind. They offer real-time project overviews, task progress, budget usage, and team workload, all in one place, and without the need to decode spreadsheets that look like a cryptographer designed them.

Plus, predictive alerts flag risks on the horizon, so you're not blindsided by missed milestones or creeping costs. You become the person who saw it coming, rather than the one explaining why it happened.

Bold Takeaway"AI won't run your project for you, but it will make you look like the kind of leader who never breaks a sweat."

Reflection Prompt What part of your current project management process feels the most like a migraine? Scheduling? Resource juggling? Progress tracking? Think about what it would feel like if that stress disappeared, what could your team accomplish with less chaos and more clarity?

AI for Customer Relationship Management – Personalization at Scale

Let's be honest: you didn't become an entrepreneur to send out mass emails that begin with "Dear [First Name]." You wanted to build something meaningful, something that makes people feel seen, even if you can't possibly remember the name of every client's dog.

That's where AI enters the chat, wearing a name tag that says: "Here to Help You Care at Scale."

A Crystal Ball for Customer Insight What if you were to have a sixth sense that tells you what your customers are thinking, before they even say it out loud. Not in a creepy, surveillance-state way, but in a "Hey, we noticed you always order extra pickles, so here's a new product you might love" kind of way.

AI-driven customer profiling tools gobble up data from everywhere, purchases, website clicks, emails, social media, and weave it into a story about what your customer values, needs, and is probably about to want. It's like having a magical assistant who reads the room... and the browser history.

Real-Time Relevance Without the Guesswork With AI personalization engines, your emails stop sounding like a corporate memo and start sounding like they were written by someone who knows your customer actually prefers oat milk.

These tools track behavior as it occurs, adjusting messages and offers in real-time. It's as if your business sent a handwritten note the moment a customer browsed a new category or paused a subscription. The kind of attentiveness that builds loyalty, and might just make someone say, "Wow, they *get* me."

Bots That Chat Like a Friend Who ListensAI-powered chatbots have come a long way since the clunky days of "I'm sorry, I didn't understand that." Today's bots remember past interactions, anticipate needs, and reply with surprising warmth. It's like having a team of customer service reps who never get tired, cranky, or stuck in traffic.

And while these bots handle the bulk, you and your team can jump in for the moments that require a human touch, meaning your energy is spent where it matters most.

Saving Customers Before They Slip Away Let's talk churn, every entrepreneur's least favorite stat. It sneaks up quietly and costs you more than you realize. But AI doesn't just wait for customers to leave. It notices subtle signals, less engagement, fewer purchases, longer response times, and sounds the alarm.

That gives you a window to act. Send a check-in message. Offer a tailored deal. Or just remind them why they loved you in the first place. With AI, you can spot the storm clouds before the thunder hits.

Feedback That Speaks VolumesAI doesn't just collect feedback; it interprets it like a wise old bartender who's heard it all. Sentiment analysis tools sift through reviews, emails, and even social posts to tell you what people *really* think.

It's not about five-star ratings, it's about spotting what's going well (double down!) and what's subtly irritating customers (fix it fast!). It's listening at scale, and that kind of listening builds businesses that last.

Bold Takeaway"AI doesn't replace your personal touch, it scales it, sharpens it, and gives it superpowers."

Reflection Prompt Think about your favorite customer. The one who loves what you do and tells their friends. Now ask yourself: how many more people *could* feel that way, if you had the time and insight to reach them just right?

From Chaos to Clarity - Your AI-Powered Business Toolkit

So here we are, at the edge of a transformation. You've now peeked under the hood of how AI can help you stop juggling flaming swords while riding a unicycle, otherwise known as "running a business."

We've toured the workhorse applications: automation that gives you your time back, dashboards that replace guesswork with clarity, communication tools that make remote teams feel like neighbors, project management that doesn't feel like herding caffeinated squirrels, and CRM tools that let you care for customers like a neighborhood shopkeeper, even if you serve thousands.

But here's the quiet truth that ties it all together: AI doesn't make your business cold, mechanical, or impersonal. It does the opposite. It handles the background noise so you can tune in to what matters, your people, your purpose, your crazy, wonderful ideas.

Entrepreneurship isn't just about efficiency. It's about creating something that feels like you. And when used wisely, AI lets you build that something faster, smarter, and with less hair-pulling and fewer sleepless nights.

So, take a breath. The heavy lifting? It doesn't have to be yours anymore. You've got AI in your corner now, loyal, tireless, and only a little too enthusiastic about spreadsheets.

And that means the next chapter of your business? It's going to be lighter, freer, and more aligned with why you started all this in the first place.

Chapter Eight

Future Trends & Emerging Technologies

The Future of AI: From Gimmick to Game Changer

LET'S SAY YOU'RE RUNNING a business ten years from now. You're sipping coffee on your balcony, laptop closed, while your AI quietly handles everything from customer follow-ups to inventory management to replying to that one overly enthusiastic client who emails at 11:47 p.m. You didn't hire an extra team, you just gave your business a brain transplant.

That's the kind of future we're steering toward: one where AI doesn't just make things faster, but smarter. Not just automated, but orchestrated.

We're entering a golden age of AI-driven personalization. In the past, "personalized" meant slapping your name at the top of an email. In the future, it means your AI remembers that you prefer early delivery windows, dislike overly chirpy customer service, and only buy red sneakers on Fridays. It's not just marketing anymore, it's memory. And it builds the kind of loyalty that no discount ever could.

What's in it for you? Predictable income, repeat customers, and less time chasing down your next sale. That's the dream, right?

Meanwhile, autonomous systems, like self-driving trucks and warehouse bots, are set to reshape the supply chain like a toddler with a box of Legos. If you're in logistics, retail, or manufacturing, these innovations will change how you staff, ship, and scale.

But don't worry, this isn't about robots replacing you. It's about finally retiring from doing all the things you hate. Manual order entry? Gone. Repetitive status checks? Done. Think of it less as "rise of the machines" and more like "fall of the menial."

Bold Takeaway: The businesses that thrive won't just use AI, they'll build *with* it.

And we're not just talking widgets and warehouses. AI is merging with the Internet of Things (IoT), which means your devices are about to get chatty. Your smart fridge won't just tell you you're out of milk; it'll nudge your AI agent to reorder oat milk because it knows you're trying to cut dairy. Your car might notice you're tired and suggest rescheduling your late meeting. Your smart office will know when to dim the lights and crank the AC before you even think about it.

For entrepreneurs, this isn't fluff. This is the path to frictionless operations and happier clients, both of which are very good for your bottom line.

And just behind the scenes? Blockchain is quietly putting on its cape and keeping your data secure. Think less buzzword, more bodyguard. With AI and blockchain paired up, you get power and protection: insights without leaks, automation without trust issues.

But as always, the magic is in the execution. In the next section, we'll talk about what *not* to do as you start integrating these technologies and how to avoid turning tomorrow's opportunity into today's PR nightmare.

Smarter Agents, Bigger Impact - The New Capabilities on the Horizon

If AI tools today are like smart assistants, good at following instructions, but not quite ready to take the wheel, then the next generation of AI agents are basically showing up with a driver's license, a suitcase, and a playlist curated to your mood.

These future agents won't just respond. They'll anticipate.

You'll be chatting with one in your CRM while it's simultaneously analyzing customer sentiment, flagging a client you forgot to follow up with, and suggesting your Tuesday schedule needs breathing room because it knows you haven't taken a real break in two weeks.

That's not science fiction. That's where things are headed.

What's in it for you? Time back. Less guesswork. And a team of virtual helpers who don't call in sick, need training, or forget deadlines.

These agents are becoming conversationally fluent, like a friend who finishes your sentences, but without the unsolicited life advice. Thanks to advancements in natural language understanding, they're starting to grasp context, tone, and nuance. You'll be able to talk to your AI in shorthand, and it'll still get what you mean.

What if an agent could detect when you're feeling overwhelmed (based on your tone, calendar load, or even typing speed), and it proactively clears low-priority tasks from your plate. It's not just helpful. It's emotionally intelligent.

And as they evolve, AI agents won't just *talk* like humans. They'll *see* and *hear* like them, too.

Using multimodal sensory integration—essentially combining inputs from text, audio, and visuals - they'll understand their environment with remarkable accuracy. In retail, an agent might use a camera to scan shelf inventory, cross-check it with sales data, and automatically reorder products. In healthcare, it might interpret X-rays and patient notes in tandem to flag potential issues.

They're becoming, in short, a second brain, with better memory.

Bold Takeaway: AI agents are moving from task-doers to thought partners. Your competition may already be giving them a seat at the table.

And the real kicker? These agents won't stay in one lane. They'll be cross-functional, handling operations, marketing, customer service, and data analysis at the same time. Think of them as Swiss Army teammates: adaptable, scalable, and oddly insightful.

A single agent might manage your email list, analyze click-through rates, recommend A/B test subject lines, and alert you that a client with high lifetime value hasn't opened your last three newsletters. That's not just efficiency. That's foresight.

Of course, more power means more pressure, especially when it comes to trust, privacy, and keeping your AI aligned with your goals. In the next section, we'll explore what it takes to build AI systems that work *for* you, not against your values or your sanity.

Building AI That Doesn't Bite

What if you finally decide to roll out an AI chatbot for your business? It's sleek. It's fast. It's going to revolutionize your customer service, or so you think. But two days in, a longtime client messages you personally to say the bot misgendered them, offered a bizarre refund policy you never approved, and somehow recommended a product you stopped selling last year. Not exactly the future you envisioned.

Welcome to the *AI backfire,* where good intentions meet poorly guided automation.

Here's the hard truth: even smart tech needs supervision. AI isn't a "set it and forget it" crockpot; it's more like a sous-chef who *thinks* they know the recipe but will happily dump cinnamon into your marinara unless you give them clear instructions.

What's in it for you? Preventing mistakes before they cost you clients, reputation, or hours of your life spent undoing robotic mayhem.

The solution? Guardrails. Not just policies or checklists, but actual frameworks that make your AI systems safer, smarter, and aligned with your brand's values.

1. Teach your AI what *not* to do.Most business owners focus on what AI *should* do: answer queries, book appointments, and analyze data. But it's equally important to define what's off-limits. For example, should your AI ever offer a discount? Transfer a customer to a live rep? Apologize? If you don't define those limits, it will make its best guess. And AI, for all its brilliance, is still just guessing based on what it's seen.

2. Human-in-the-loop is not a sign of failure; it's insurance. Let's say your AI drafts a contract. It's 90% correct. But that 10% can be a BIG problem. That's the part that might accidentally waive a fee, promise overnight shipping to Guam, or leave out a non-compete clause. By adding a human review stage, especially for high-stakes or client-facing outputs, you protect the business and preserve your voice.

Bold Takeaway: "AI is powerful, but it's not magic. If you don't tell it what matters, it'll make assumptions, and not always good ones."

3. Make ethics a design feature, not an afterthought. Ethical AI doesn't just mean "don't be evil." It means training your AI on data that reflects your values, not just your metrics. Let's say your support bot is trained mostly on complaint data. It might start assuming every customer is angry. Now your friendly family brand sounds like it's bracing for a fight. That's not alignment. That's an algorithmic mood swing.

Instead, think of ethics like customer service: It's not just there when things go wrong. It's baked into every interaction, every response, every choice the system makes on your behalf.

You don't need a PhD in moral philosophy. But you do need to ask one simple question: "If this AI decision showed up on the front page of the newspaper tomorrow, would I be proud of it?"

4. Use explainable AI tools when possible. Ever asked a chatbot "Why did you say that?" and it just... didn't answer? That's the black-box problem. Tools that offer *explainability*, meaning they can tell you what they based a decision on, are essential if you want to audit outcomes, troubleshoot mistakes, or stay legally compliant.

What this means for entrepreneurs: You don't need to be perfect, but you *do* need to be intentional. The more you align your AI with your goals, tone, and values, the less likely it is to turn into a rogue digital intern with attitude.

Next up, we'll wade into the privacy swamp, don't worry, we've got a map and a good pair of boots.

Keeping Your AI Compliant (Without Losing Your Mind)

Most entrepreneurs would rather eat glass than read a privacy policy. Honestly, I'm not sure I'd blame them. Terms like GDPR, CCPA, and "data sovereignty" tend to sound less like business tools and more like the names of distant planets you hope never to visit.

But here's the thing: your AI can't play fast and loose with people's information, *not anymore*. And no, you don't need to be a lawyer. You just need to understand a few key principles that keep you (and your tech) out of trouble.

What's in it for you? You sleep better. You avoid fines. You don't wake up to a PR nightmare trending on X (formerly known as the hellscape of Twitter).

Bold Takeaway: "Privacy isn't just legal, it's a selling point."

Customers care. Regulators care. And if you ignore this stuff, your competitors will eat your lunch by saying, "Unlike others, we *actually* respect your data."

So, here's your no-jargon cheat sheet to not getting sued:

1. Data Minimization: Just because you can collect it, doesn't mean you should. Let's say you run an e-commerce store. You don't need someone's birthday, pet's name, or blood type just to ship them socks. Collect only what's necessary, and store it for only as long as you need. It's like dieting for data. Trim the fat, keep the essentials.

And if your AI tool is sucking up every scrap of information like a toddler with a juice box, it's time to put some limits in place.

2. Consent: If it's not clear, it's not consent. You can't hide data collection in 17 pages of fine print anymore. People need to know *what* you're collecting, *why*, and *how it'll be used*. And no, a tiny checkbox under a 3-point font pop-up doesn't cut it.

Make it obvious. Give them a real choice. And don't punish them if they say no.

Bonus tip: The more transparent you are, the more trust you build. People appreciate being asked, even when they say no.

3. Keep Thing Private and Secure It's one of those things that seems obvious in hindsight. Privacy should be *built in* from the start, not duct-taped on at the end like a sad disclaimer.

Using AI? Make sure your vendors follow good security practices. Don't store sensitive info in plain text. Don't train models on customer data without permission. And whatever you do, don't email yourself a spreadsheet full of passwords. (Yes, people still do this.)

Think of this as designing a car with seatbelts *before* you crash.

Entrepreneur Lens: This isn't about red tape. it's about protecting the thing you've worked so hard to build. Your brand. Your customer relationships. Your reputation. The goal isn't just to follow the law. It's to stay ahead of it and build a business people can actually *trust*.

Now that you've got your privacy pants on, let's shift gears and talk about something even more slippery: bias.

Next stop, how to be a better human *and* a smarter entrepreneur, all while making sure your AI doesn't accidentally offend half your customer base.

Bias, Inclusion, and Being a Good Human

A while back, a startup launched an AI-powered resume screener that was supposed to help them hire faster. Instead, it systematically weeded out qualified candidates who didn't fit the same mold as their current team. It wasn't malicious; it was just trained on past data. But past data, as it turns out, carries all sorts of baggage. What they got was not a smarter hiring tool, but a digital echo chamber.

Here's the kicker: when they caught the bias and fixed it, something surprising happened. Not only did they get better hires, they also earned praise from customers and partners who appreciated the transparency.

What's in it for you? Better decisions, wider reach, and fewer moments that end with "we sincerely apologize for the oversight."

Bold takeaway: "Inclusive AI doesn't just avoid lawsuits, it builds loyal fans."

So how do you keep your tech from turning into a well-meaning but wildly inappropriate dinner guest?

1. What Bias Actually Looks Like

Bias in AI doesn't usually show up wearing a big neon sign. It's sneakier than that. Maybe your chatbot responds more helpfully to one type of name than another. Maybe your pricing model favors ZIP codes that, statistically, correlate with higher income, and less diversity. Maybe your product recommendations look suspiciously different for men than women.

Bias isn't just offensive. It's expensive. It means lost customers, missed markets, and brand damage that can take years to undo.

2. Inclusive Design: Not Just Morality - Strategy

Let's be clear: designing for inclusion isn't charity. It's business smarts.

If your app can be used equally well by people of all backgrounds, languages, and abilities, you open the door to more customers, more referrals, and more love from regulators. You also get feedback from a broader pool, which means your products improve faster and serve more people.

And yes, customers *notice*. In fact, they remember the businesses that make them feel seen.

3. The Business Case for Diverse Input

The more varied the people in the room, whether literally or metaphorically, the better your AI gets.

If your dev team all went to the same school and binge the same shows, guess what your AI's going to reflect? A narrow worldview. And in business, a narrow worldview is a liability.

Invite diverse testers. Get feedback from people outside your bubble. Use open datasets that represent a range of voices and experiences. Think of it as upgrading your AI's worldview from "suburban cul-de-sac" to "Times Square."

Reflection Prompt:"Who's not being served by your current system, and how would you know?"

If your AI had blind spots, would you even catch them? What questions aren't being asked? What customers aren't being heard?

The answers could save you time, reputation, and a whole lot of awkward explaining later.

Tech You Can Trust, Business You Can Sleep With

So here we are, at the end of our little guided tour through the minefield of AI mishaps. Along the way, we've seen how one misfired chatbot can tank your reputation faster than you can say "Oops, that wasn't supposed to happen." We've peeked at compliance without developing a rash, unpacked the quiet chaos of algorithmic bias, and discovered that your AI should probably pass the "would-I-trust-this-thing-to-talk-to-my-mother?" test before going live.

But let's zoom out for a moment.

At its core, this chapter wasn't about code or compliance, it was about **trust**.

Not the vague, squishy kind of trust that comes from handshake deals and networking lunches, but the very real trust that underpins your business relationships. Your brand, your team, your customers, they're all relying on you to make choices that don't just scale *fast*, but scale *right*.

What's in it for you? Peace of mind. A solid reputation. And a business that doesn't accidentally punch its customers in the face while trying to shake their hand.

Bold takeaway: "AI should scale your values, not replace your judgment."

You don't need to be an ethicist or a philosopher to use AI responsibly. You just need to lead with curiosity, empathy, and a good dose of common sense. And maybe a little paranoia when your automated emails start sounding suspiciously like a Bond villain.

Here's the real trick: Use AI to do what it does best, handle the repetitive, the routine, the riddled-with-data stuff, so you can double down on what you do best: leading with heart, vision, and human smarts.

Final image: Picture the AI system that works like your favorite employee. The one who never calls in sick, doesn't mind the boring tasks, and somehow always remembers the client's dog's name. It's fast, it's smart, and, most importantly, it's aligned with your mission. That's the kind of tech you can build. That's the kind of business you can believe in.

And maybe, just maybe, sleep soundly knowing your digital team isn't accidentally burning the house down while you're dreaming about quarterly goals and beach vacations.

Ready to step into the future with your values intact and your reputation polished to a shine?

Then buckle up, because the next chapter's where it gets really interesting.

Chapter Nine

Building a Scalable AI Strategy

Designing AI Solutions for Scalability

IMAGINE PLANNING A WEEKEND getaway. You pack flip-flops, a book you'll pretend to read, and that one shirt that always looks good in vacation selfies. But just as you zip your bag, you remember that rain is always a possibility. So, in goes the emergency umbrella and the hoodie with questionable stains. That's scalable AI in a nutshell, preparing for sunshine, downpours, and surprise picnics on the beach of business chaos.

When you're building AI for your company, you're not just solving problems for today. You're packing for growth, pivots, staff changes, product expansion, and whatever curveball the market throws next. That means your tech needs to stretch without snapping and help you stay sane in the process.

What's in it for you? You avoid rebuilding everything from scratch six months down the line, your tools keep pace with your growth, and you get to say, "We planned for this" instead of "Why is everything on fire?"

Let's break it down.

Step One: Identify the Bottlenecks

Start with the parts of your business that feel like wading through molasses. Customer service queues that never end? Manual data entry that requires caffeine and therapy? Those are prime candidates for AI to step in, not as a robot overlord, but as your business's favorite intern (minus the need for lunch breaks or Spotify).

Think automation first: Where is time wasted? Where are human brains doing what machines could do better? The goal isn't to replace your team, it's to make them unstoppable.

Bold takeaway: If it's boring, repetitive, or error-prone, AI can probably do it faster, cheaper, and without asking for a raise.

Step Two: Think Bigger (Without Building Bigger)

Once you've handled today's headaches, start squinting into the future. Where are you going in 6 months? A year? Do you want to expand into new markets, launch new products, or serve ten times as many clients without losing your mind?

Scalable AI means designing systems that grow with you. The magic phrase here is **modular architecture**. Picture Lego blocks: you build what you need now, but later you can snap on new parts, like analytics, language tools, or a chatbot that doesn't sound like a confused squirrel.

What's in it for you? Flexibility. When the market changes (and it will), you won't be stuck duct-taping a Frankenstein system together at 2 a.m.

Step Three: Use the Cloud (Unless You Like Server Rooms)

Unless you're nostalgic for the sound of humming servers and the smell of burning coffee near a desktop tower, you'll want your AI hosted in the cloud. Cloud platforms give you power on demand, scale up when things boom, scale down when they don't.

It's like having an electric bill that shrinks when you turn off the lights. Only in this case, you're turning off a thousand virtual processors.

Bold takeaway: The cloud keeps your AI lean, fast, and ready to flex with your business, not anchored to hardware that ages faster than an avocado.

Data Scalability: Build Bigger Pipes

As your AI gets smarter, it's going to want more data, like a teenager going through a growth spurt and discovering carbs. If your systems can't handle the flood, everything slows down. That's where **distributed data storage** comes in: sharding, partitioning, and other geeky tricks to help your data flow smoothly, even under pressure.

What's in it for you? Your systems stay fast and responsive even when you hit it big, and you'll spend less time watching spinning wheels and more time watching profits.

Adapting AI to Evolving Business Needs

Running a business these days is like dancing at a wedding where the DJ changes the music every 30 seconds. One minute it's a slow waltz, the next it's full-blown techno, and then, for some reason, there's a sudden bagpipe solo. If your AI can't keep up with the tempo, you'll be stepping on toes, or worse, sitting out the dance entirely.

Adaptability isn't a "nice to have." It's a survival skill. Markets shift. Customer preferences do somersaults. And that tool you installed last year? It might already be outdated. AI that can't evolve will age out faster than a Blockbuster card.

What's in it for you? You stay flexible, fast, and focused, able to pivot without panic, and lead without needing to reinvent everything from scratch.

Routine Checkups for Your AI

Your AI strategy needs the same kind of checkup your doctor keeps nagging you about. Regular assessments help you spot when something's off, or when there's a new opportunity staring you in the face. Maybe a product is suddenly hot in a new market. Maybe customers keep asking questions you didn't think to train your AI Agent for.

AI reviews should be baked into your calendar. Think of them as quarterly tune-ups for your business brain.

Bold takeaway: Don't just set it and forget it. Schedule time to reassess where your AI can help you win the next round, not just survive the current one.

Build Systems That Can Pivot

The real secret to staying relevant is building AI that can turn on a dime. Enter: **adaptive algorithms** and **flexible integrations,** the Swiss Army knives of modern automation. These tools can shift with your business, like a chameleon that also balances your books.

Make sure your systems can plug into new tools and talk to new platforms without a six-month IT project. Your AI should be a good conversationalist, not a hermit locked in a legacy database.

What's in it for you? No matter what changes, your tools change with you. No breakdowns. No panic. Just smooth transitions and smarter moves.

Collaboration: Not Just for Brainstorming

Here's the part no one tells you: adapting AI isn't just about tech, it's about teamwork. Your marketing team, sales team, and tech folks all see the world differently. And that's good. Because when you bring them together to plan your AI strategy, sparks fly in the best way.

Hold cross-department workshops. Form AI tiger teams. Treat it like a company potluck, where everyone brings a dish, and no one is allowed to bring plain salad.

Bold takeaway: The best AI ideas don't come from one department; they come from the places your org chart forgot to connect.

Keep a Pulse on the Market

Want to stay ahead? You've got to watch the waves before they hit. Use tools for real-time market analysis. Set up alerts for trends in your industry. Follow your competitors like you're checking on your ex's new relationship (but, you know, professionally).

When you build trend forecasting into your AI plans, you go from reactive to proactive. You don't just follow the market; you dance ahead of it.

What's in it for you? You spot opportunities before your competitors do, and dodge mistakes before they become expensive PR disasters.

Building an Agile AI Framework for Continuous Improvement

If your AI strategy is a once-a-year meeting where someone says, "Didn't we install something last March?", you're doing it wrong. In the world of modern entrepreneurship, AI isn't a statue you put on a shelf and admire. It's more like a sourdough starter, constantly growing, a bit unpredictable, and in need of regular feeding if you want it to perform.

The trick is to stay flexible. Not yoga-flexible, agile-flexible. That means building AI systems that evolve with your business instead of rigidly marching off into irrelevance.

Think in Sprints, Not Monuments

Forget building the Parthenon. Think Lego kits. Break big projects into small, testable pieces. Agile methodology turns AI development into an ongoing conversation instead of a one-time declaration. You try something. You see how it works. You adjust. Repeat.

This approach saves you from the classic trap of spending six months (and a bucket of money) building something that no one wants by the time it's finished.

Bold takeaway: Agile AI is better AI. It lets you fail faster, learn sooner, and win more often, without burning down the house to rebuild the kitchen.

Feedback Loops: The Lifeblood of Better AI

Ever served dinner to a toddler? You get instant feedback. Same goes for AI. You need built-in ways to listen, learn, and improve, ideally without the screaming.

Collect user feedback. Run internal audits. Watch how the AI performs in the wild. Then tweak it like a master chef perfecting a secret sauce.

Your AI is only as good as the last round of improvements. So, keep the loop open.

What's in it for you? You catch small issues before they become big messes. And you show your customers you're actually listening, something your competitors probably aren't.

Experiment Like It's a Science Fair

Encourage your team to test weird ideas. Not all of them will work. That's the point. Pilots and prototypes let you explore risky terrain without betting the farm.

And when something flops? Celebrate it. You just saved yourself a major faceplant by catching the issue while it was still small. (Bonus points if you learned something that leads to the next breakthrough.)

Bold takeaway: Innovation doesn't come from avoiding mistakes. It comes from making the right kind of mistakes in a smart, controlled way.

Never Stop Learning - AI Sure Doesn't

AI evolves faster than a toddler on sugar. If your team's knowledge stays frozen in time, your tools will too. Continuous learning isn't an optional extra, it's the fuel for staying relevant.

That means AI training, workshops, deep dives into emerging trends, and (let's be honest) a few YouTube rabbit holes. Equip your people to grow with the tech, and they'll help your business grow in return.

What's in it for you? A team that stays sharp, nimble, and ready to pivot when opportunity knocks—or when the market shifts under your feet.

Aligning AI Strategy with Business Goals and Vision

Let's imagine you're about to set sail on the voyage of a lifetime. You've got a shiny new ship (AI), a full crew (your team), and enough enthusiasm to power a small wind farm. One small problem: no one packed a map. You're not entirely sure where you're going, you just heard "AI" is where the future is headed, and figured you'd better hop on board.

And this, dear reader, is how a lot of businesses end up lost at sea with a very expensive robot that doesn't do what they hoped.

Let's fix that.

Strategy First, Then Technology

AI is not the goal. It's a tool. A means. Like a corkscrew or a pressure washer, it's great, but only if you know what you're trying to open or clean up.

Before plugging anything in or launching fancy dashboards, get crystal clear on what matters to your business right now and where you want it to go. Are you trying to reach new customers? Speed up fulfillment? Reduce support costs? Fine. But pick a lane. Then make sure your AI is pointed in that direction.

Bold takeaway: The worst AI strategy is one that's "cool" but completely disconnected from what your business actually needs.

Strategic Alignment Isn't Just Jargon

To make your AI investment count, it needs to be married, happily, to your business goals. That means looping in leadership, syncing with department heads, and making sure everyone's rowing the same direction.

Run a strategic alignment session. Ask: "How can AI help us reach this quarter's top priorities?" Not "What's the fanciest tool we can afford?" Align incentives. Track outcomes. Repeat quarterly.

What's in it for you? You stop wasting time and money on tools no one uses and start building systems that actually drive results.

Communicate the Vision, Repeatedly

This is the part where a lot of good ideas die quietly in a Google Doc. If your team doesn't understand the AI vision, or worse, fears it, they won't use the tools, and your fancy automation ends up gathering dust.

You need clear messaging. Why are we doing this? How will it help our customers? What's in it for the team? Say it in plain English. Say it often. Then say it again, maybe with donuts this time.

Create space for feedback, too. Good ideas bubble up from the front lines, not just the C-suite.

Bold takeaway: People won't get behind your AI strategy unless they understand it, believe in it, and see where they fit into the story.

Review and Realign (Quarterly is Not Overkill)

Even the best strategies need tune-ups. Business priorities shift. Markets wobble. Something new explodes on TikTok. Whatever the reason, your AI needs to flex with it.

I know, I've said to review quarterly already, but some things merit repeating. Set quarterly check-ins. Revisit what's working. Drop what's not. Update your goals and re-anchor your systems. It's less about chasing perfection, more about staying relevant.

What's in it for you? You avoid the slow fade into irrelevance that happens when AI runs on autopilot, and no one's checked the compass in six months.

Final Image: AI as a First Mate

When your AI strategy is truly aligned with your business goals, it doesn't feel like some high-maintenance diva in the corner of the office. It feels like your most reliable crew member, smart, dependable, always on duty, and pulling in the same direction you are.

And just like a great first mate, it makes sure you're not sailing blind into stormy waters, or worse, in circles.

Wrap-Up: From Paper Plans to AI with Purpose

So here we are at the end of our little strategic stroll, hopefully with fewer blisters and a clearer sense of where we're going. If there's one thing to carry with you from this chapter (besides a vague craving for a picnic), it's this: building scalable AI isn't about throwing tech at a wall and seeing what sticks. It's about intention. Alignment. And maybe, just maybe, packing your metaphorical bags with more than just good intentions and wild optimism.

Scalable AI is the kind that grows with you, like a faithful hound that learns your route and eventually starts carrying the map. You're not just solving today's problem, you're building an adaptable foundation for tomorrow's expansion, evolution, or midlife business pivot.

That means designing with flexibility, planning for growth, and aligning every AI move with your business strategy, not because it sounds impressive in a boardroom, but because it works better that way.

It means creating systems that can flex, evolve, and surprise you in good ways, and knowing when to step in and adjust when they don't.

It means having feedback loops, cross-team collaboration, and regular check-ins, not because you love meetings, but because you love progress.

And above all, it means keeping your eye on the real win: a business that runs smarter, not just faster. One where you're in the captain's seat, not duct-taped to the rudder while a rogue AI charts a course toward the Bermuda Triangle.

So, as you put this chapter to rest and plot your next steps, remember: AI isn't magic. But used well, it can feel like magic.

Now let's take all this strategic goodness and put it to work where it really shines, your customer experience. Onward!

Chapter Ten

Creating a Culture of Continuous AI Learning

The Learning Buffet (and How Not to Get Indigestion)

THINK ABOUT STEPPING INTO an AI knowledge buffet, not the kind with sneeze guards and questionable crab dip, but a glorious spread of courses, books, and insights. The table stretches on forever, with flavors ranging from "Intro to Machine Learning" to "How Neural Networks Think (Sort Of)." As an entrepreneur, you don't have time to taste everything. You're trying to run a business, not earn a PhD. The real challenge isn't access, it's discernment.

And here's the rub: when it comes to AI, ignorance isn't bliss. It's expensive. Whether it's choosing the wrong tool, misinterpreting what an algorithm does, or simply zoning out while your competitors quietly automate their way into your market share, the cost of not learning stacks up fast.

What's In It For You?You don't need to become a machine learning wizard. But you do need to speak the language, understand the basics, and stay sharp enough to make smart decisions—or at least know when to ask better questions., Let's start with the best of the buffet. Online platforms like Coursera, edX, and Udacity serve up university-level AI courses in a format that fits between morning coffee and your third meeting of the day. Courses like "AI for Everyone" by Andrew Ng break things down without melting your brain. These are your Michelin-starred learning stops, zero fluff, all flavor.

But don't underestimate good, old-fashioned books. Titles like *Artificial Intelligence: A Guide to Intelligent Systems, Rebooting AI, or Agent AI for Entrepreneurs (o.k., was that self-serving)* are the nonfiction equivalent of a long, thoughtful dinner, slow to digest, yes, but they stick with you. Books let you pause, reread, dog-ear, and ponder. They're your AI slow-cooker: deeper flavors, lasting impact.

Bold Takeaway: "AI learning isn't about mastering the code. It's about building the confidence to lead, not follow."

Join the Conversation (Even If You Don't Know What to Say Yet)

If self-paced courses are the study hall of AI learning, communities are the after-hours club where the real stories get shared. Reddit's r/MachineLearning might sound intimidating, but you'll find a healthy mix of newbies, skeptics, and tinkerers asking questions like, "Why is my chatbot sending love poems to customers?" (True story.)

LinkedIn also hosts surprisingly lively AI groups, which feel like a water cooler meeting a TED Talk. Joining these groups does two things: it keeps you informed, and it builds relationships with people who are a few steps ahead of you. The kind of folks who've already fallen into the potholes you haven't hit yet.

Bold Takeaway: "Lifelong learning in AI isn't solo work. It's a team sport with an open bench."

Next Up: From Curiosity to Culture

Now that you've found the learning feast and figured out how not to drown in it, the next step is bringing that hunger for growth back to your business. What if your whole team were on the same AI-empowered page? What if learning didn't stop with you?

That's where we're headed next: how to build an organization that learns and grows with AI, together.

Developing Internal Learning Initiatives

Turning Your Business Into a Brain Trust (Without the Hogwarts Robes)

Let's suppose you've dipped your toe into the AI learning waters and found the temperature rather agreeable. Maybe you've even earned a certificate or two, shared a few insightful LinkedIn posts, and started recognizing terms like "neural net" without your pupils dilating in fear. Good. Now, imagine what would happen if your *entire team* got on board.

You see, learning can't just live in the founder's brain. That's like owning the only flashlight in a cave and expecting everyone else to find their way by sheer optimism. If you want AI to transform your business, it has to become part of your company's operating rhythm, like meetings, coffee, and that one guy in accounting who always brings tuna for lunch.

So, how do you bake learning into your culture without turning your business into a lecture hall?

Start With Internal Workshops (That People Actually Want to Attend) If you have a company that's too small for an internal workshop, find good articles to share, or even initial a spirited conversation occasionally. Think bite-sized sessions that answer real questions: What's AI capable of right now? What tools are already on our computers that we're underusing? Could AI help shorten our sales cycle? (Spoiler: Yes.) Make it engaging. Feed people. Seriously, feed them. Learning retention increases dramatically when there are snacks involved, science probably says so somewhere.

Mentorship: Because Learning Shouldn't Be Lonely Pair up the curious with the confident. Maybe your marketing director has been playing around with generative AI, while your operations lead still thinks "GPT" is a type of motor oil. Let the experienced folks guide the eager ones. It's good for morale, good for knowledge sharing, and great for reminding your internal experts that they actually *are* experts.

And for those of you who are solo entrepreneurs or running very lean teams, this still applies. Look around at your contractors, partners, or even mastermind groups. Learning doesn't have to be confined to payroll.

Bold Takeaway: "When your whole team learns together, you don't just get smarter, you build shared momentum."

Let the AI Teach the Humans (Irony Fully Intended) Adaptive learning platforms, those clever programs that adjust to your progress, are like the Peloton of knowledge. They push when you're ready, slow down when you're not, and gently shame you when you stop showing up. Think platforms like Degreed, Coursera for Business, or LearnUpon. These tools give your team personalized paths to AI literacy without you having to map it all out.

And better yet? The same tech you're learning *about* is the tech helping you learn it. That's practically poetic.

Progress Isn't a Feeling, It's a Feedback Loop You wouldn't run a business on vibes alone (hopefully), so why treat learning that way? AI-powered learning tools give you real-time dashboards that show who's engaged, who's stuck, and who might need a nudge. You can track growth in actual skill sets, not just who seemed enthusiastic at the lunch-and-learn.

This isn't just good HR,it's smart strategy. It helps you align training with goals and spot the emerging talent who might become your future AI lead.

Bold Takeaway: "The real magic happens when learning becomes measurable, social, and baked into your everyday operations."

Fostering Innovation Through Collaborative AI Efforts

Where Cross-Pollination Meets the Avengers (Minus the Capes)

Think about your business not as a machine, but as a rainforest. Every team, from marketing and sales to IT and operations, is its own quirky species, each contributing to the ecosystem in a beautifully chaotic way. Now, left alone, these teams might survive just fine. But bring them together in the right conditions, and suddenly... innovation blooms like an orchid in the wild.

When it comes to AI, this sort of cross-pollination isn't just nice, it's necessary. The best AI breakthroughs rarely come from a single desk or department. They come from unlikely conversations between someone who understands the customer, someone who knows the numbers, and someone who can write Python like it's poetry.

The Innovation Team You Didn't Know You Needed Start small. A cross-functional AI team might sound fancy, but it can be as simple as pulling together one curious person from each department. You don't need a PhD from MIT. What you need is a group of people who ask good questions, aren't afraid to fail, and can explain what they're trying to solve in plain English.

Give them a challenge, maybe customer churn, maybe operational inefficiency—and let them play. Give them time, space, and support. You might be amazed by what comes out of a few weeks of structured creativity.

Hackathons: Because Nothing Fuels Genius Like a Little Chaos Here's the scene: two days, a whiteboard, a pile of snacks, and one question: What could we build if we had no red tape? That's the spirit of an internal AI hackathon. It's part jam session, part team-building retreat, part lightning storm of ideas.

These aren't just feel-good exercises. They can be testing grounds for the next tool that saves you six figures or shaves weeks off your sales cycle. And let's be honest, your team will remember the thrill of creating something real far more than any memo you've ever sent.

Bold Takeaway: "Innovation doesn't always come from the top; it often comes from the team that was given permission to experiment."

When You Don't Know, Borrow Genius Now here's a secret every seasoned entrepreneur learns eventually: You don't have to know everything, you just have to know where to find someone who does (wasn't that Henry Ford's superpower?). That's where partnerships come in.

Collaborating with AI startups, universities, or research labs gives you access to cutting-edge thinking without having to reinvent the algorithmic wheel yourself. Many of these organizations are actively looking for small businesses to test their latest models or partner on pilot programs. You bring real-world problems; they bring bleeding-edge tech. It's a match made in entrepreneurial heaven.

And let's not forget the talent pipeline. Tapping into university programs can bring in hungry, creative minds who haven't yet been told something "Can't be done."

Create a Playground, Not a FortressA culture of innovation only thrives when ideas can move freely. That means ditching the "not-my-department" attitude and replacing it with something far more powerful: shared ownership.

Set up digital idea boards or feedback loops where employees at every level can pitch AI concepts, offer improvements, or simply ask curious questions. Reward the bold thinkers. Celebrate the weird ideas. And if someone suggests an AI-powered coffee machine that remembers your favorite brew? Maybe build it, just for fun.

Bold Takeaway: "The businesses that innovate the fastest are the ones where everyone feels safe to suggest the impossible."

And Finally, Use the Tools You're Bragging About Don't just talk about AI; use it to drive collaboration. Platforms like Slack, Notion, or Microsoft Teams now come packed with AI features that help teams brainstorm, draft, and organize faster. Shared notebooks, live collaboration on AI models, and automated feedback loops; these aren't just for big tech companies anymore. They're your new whiteboard, sticky note, and project tracker rolled into one.

Think of it as turning your digital workspace into an AI-enabled idea greenhouse, where every good idea gets a little sunshine and room to grow.

Staying Informed: How to Keep Up with AI Developments

How to Keep Pace Without Losing Your Mind (or Your Lunch Break)

Staying current with AI trends can feel a bit like trying to read a book during a windstorm; pages flying everywhere, and you're just hoping something sticks. But here's the good news: you don't have to be glued to tech journals or code forums at 2 a.m. to keep your edge. With a bit of intentional curation, you can stay smart on AI without making it your second job.

Let the Smart People Do the Heavy Lifting Social media isn't just for vacation photos and memes about forgetting your password again. It's where AI thought leaders, people who are neck-deep in the bleeding edge, often share their latest insights in bite-sized, digestible nuggets.

Twitter (now X, depending on your level of digital optimism) is an especially rich source. Follow AI researchers, startup founders, and technical leads. They'll post about breakthroughs, share code snippets, and occasionally make jokes only a neural network would fully understand. The point is, they're already reading the journals, so you don't have to.

And yes, LinkedIn is still relevant, especially for conversations around how AI is actually being used in business. Want to know what your competitors are tinkering with? Follow the right folks, and you'll get a front-row seat to their experiments (and their misfires).

Bold Takeaway: "You don't need to be the loudest voice in AI, just the most observant."

Inbox Gold: The Rise of the Expert Newsletter You know that warm feeling when a friend sends you a really good article, saving you from the abyss of clickbait? That's what a great AI newsletter feels like. It's curated wisdom, neatly wrapped and delivered to your inbox without any effort on your part.

Some newsletters are technical. Others are strategy-focused. A few are funny and human, making you feel like you're not the only one wondering if ChatGPT can write a decent dinner invitation. Pick two or three that speak your language and read them religiously. (Well, casually religiously, no altar required.)

Conference Magic: Without the AirfareAI conferences used to mean expensive flights, bad coffee, and jet lag. Now? Many of them are available online, often for free or cheap. That means you can learn from MIT researchers, OpenAI leaders, or the head of AI at Google without changing out of your house slippers.

Look out for big-name events like NeurIPS, AI Expo, or even TED-style AI talks. But also keep an eye out for niche webinars specific to your industry, like AI in retail, finance, or real estate. These are gold mines for insight and networking without needing to shake 300 hands.

Bold Takeaway: "You don't need a lanyard to level up."

The Entrepreneur's Reading List For those moments when you do want to dig deeper, say, during a flight or a Sunday morning with coffee, turn to books and long-form journalism. No, you don't have to slog through textbooks. There are smart, readable books that translate complex AI into something that doesn't feel like you're being held hostage by an algorithm.

And if you're more of a podcast person? Perfect. AI-themed shows are popping up faster than you can say "machine learning." Listen while you walk the dog or do the dishes. It all counts.

Your AI Assistant, Curating AI News Irony alert: you can actually use AI to stay on top of AI. Tools like Feedly or Flipboard let you build custom dashboards that pull in only the most relevant stories. No more scrolling through 27 articles on AI-powered toasters before you find something useful. You train the tool once, and it becomes your personal headline concierge.

Combine that with Google Alerts for your favorite keywords, and voilà, you've got an automated stream of relevance without drowning in fluff.

The Real Goal: Integration, Not Obsession At the end of the day, the point isn't to be an AI encyclopedia. It's to stay curious, stay open, and find smart ways to apply what you learn to your business. Read enough to ask better questions. Follow enough to know what's possible. That alone will put you miles ahead of entrepreneurs who are still trying to figure out what AI actually is.

Embracing Change: Cultivating an Adaptive Entrepreneurial Mindset

Wearing Your Cape with Confidence

If you've ever tried updating your phone's software only to find yourself wishing for a rotary dial again, you already understand the emotional range that comes with change. Now magnify that to the scale of running a business in the age of artificial intelligence, and you'll see why cultivating an adaptive mindset is no longer optional; it's your secret sauce.

Think of it this way: you wouldn't send a sailor to sea without teaching them how to tack with the wind. In the same spirit, you can't lead a business into an AI-powered future without embracing the art of flexibility and the magic of a good pivot.

Growth Mindset: The Brain's Stretchy Pants Creating a growth mindset isn't about reading motivational quotes until your eyes glaze over. It's about rewiring how you and your team respond to hiccups. When something breaks (and trust me, something will), the goal isn't to avoid the fall, it's to land like a cat and saunter off as if you meant to do that.

Hold regular workshops or training sessions that encourage your team to try, fail, learn, and try again, with enthusiasm instead of dread. The workplace equivalent of a mental gym. No barbells required.

Bold Takeaway: "Challenges aren't roadblocks, they're launchpads with identity issues."

The Quiet Superpower: Reflective Practices In the rush to get things done, reflection can feel like a luxury. But here's a secret: it's one of your best productivity tools in disguise. Build in time for you and your team to look back, ask what worked, what flopped, and what might be worth tweaking. Like a GPS that recalculates after every wrong turn, reflection doesn't judge, it redirects.

Whether you journal after big decisions or hold quick "lessons learned" huddles post-project, make it a habit. It creates a culture where everyone's learning out loud and nobody's afraid of the occasional wrong note.

Planning for a Future That Hasn't Been Invented Yet Scenario planning might sound like something reserved for boardrooms full of suits and flowcharts, but it's actually a creative exercise disguised as strategy. You're not trying to guess the future; you're sketching a few potential versions of it so you're not blindsided when one of them shows up at your door.

What happens if a competitor launches an AI tool tomorrow? Or if your favorite platform changes its pricing model? By playing out a few "what ifs," you turn panic into preparation, and strategy into storytelling.

Bold Takeaway: "The best way to predict the future is to plot three versions of it and befriend the one that shows up."

Resilience Training: Because Business Is a Contact Sport If change is the game, then resilience is your playbook. Equip your team with the tools to bounce back, adapt quickly, and keep their cool even when the tech misbehaves (which, let's be honest, it will). Host workshops, bring in guest speakers, share real-world stories of people who stumbled but stood taller afterward.

Encourage risk-taking in ways that don't end in disaster. Start small. Launch pilot projects. Let people experiment, then regroup and refine. It's the difference between doing science and doing guesswork with fancier spreadsheets.

Failing Fast Without Falling Apart"Fail fast" has become such a trendy phrase that people forget it's not about loving failure, it's about learning quickly. Give your team permission to build things that don't work, as long as they're learning faster than they're burning out. If one idea fizzles, dust it off, learn what went wrong, and move on like a chef who's just slightly overcooked the risotto. There's always another batch.

Personal Goals Meet Company Goals: A Harmony Worth Tuning The future of work isn't just AI-powered, it's alignment-powered. That means your people grow as your company grows. Fold AI education into personal development plans. Give folks a reason to care and a path to advance.

Offer incentives for upskilling. Celebrate those who reach new milestones, not just the ones who close big deals. When your team feels like they're growing *with* the business instead of getting dragged along behind it, everything changes. Culture. Morale. Momentum.

Bold Takeaway: "You don't build a forward-thinking company by dragging people into the future, you invite them to co-create it."

If lifelong learning is the compass, then an adaptive mindset is the engine that keeps it all humming along, even when the road ahead gets bumpy. AI isn't a thing you conquer; it's a wave you learn to ride. And this mindset? It's your surfboard.

Now, as we close out this chapter and this book, keep in mind that you're not just preparing to adopt AI, you're setting the tone for a culture that thrives because of it. One more thing, as you close this book, hopefully with a head full of ideas and a few less "busy" things on your to-do list, I have one last favor to ask (and I promise, it doesn't involve spreadsheets). If you found this book helpful, enlightening, or even mildly entertaining, I'd be grateful if you'd return to Amazon and leave an honest review. Hopefully five stars, but hey, I'll take sincerity over flattery any day.

And if you'd like to be the first to know when the next book in this series makes its grand entrance, or check out one of my previous books, follow me on Amazon at . There's more to come—more tools, more stories, and more ways to help you build a business that doesn't run your life. The next book in this series; **The 30-Day AI Agent Blueprint: A Tactical Blueprint to Build, Scale, and Succeed with Autonomous Agents in Business**

Thanks for spending some time with me. Now go reclaim your day.

References

- *AI agents vs traditional automation: Understanding the key ...* https://www.geeks.ltd/insights/articles/ai-agents-vs-traditional-automation#:~:text=Traditional%20automation%3A%20Rule%2Dbased%2C,real%2Dtime%20based%20on%20data.
- *Machine Learning (ML) in Business: Automation and More* https://www.artsyltech.com/blog/Machine-Learning-in-Business
- *AI-driven customer segmentation: A step-by-step guide* https://www.revenuemarketingalliance.com/ai-driven-customer-segmentation/
- *The 19 Best AI Tools for Small Businesses and How to Use ...* https://blog.hubspot.com/sales/small-business-ai-tools
- *AI Readiness Assessment - Learn Microsoft* https://learn.microsoft.com/en-us/assessments/94f1c697-9ba7-4d47-ad83-7c6bd94b1505/#:~:text=An%20AI%20Readiness%20Assessment%20evaluates,Management%2C%20and%20Azure%20Open%20AI.
- *AI Integration Challenges: Insights for Competitive Edge - Aura* https://blog.getaura.ai/ai-integration-challenges

- *The 8 Best AI Tools for Entrepreneurs & Startups in 2025* https://altar.io/the-best-ai-tools-for-entrepreneurs-startups/
- *AI For Business - 30 Case Studies That Led To Competitive ...* https://digitaltransformationskills.com/ai-for-business/#:~:text=JPMorgan%20Chase%20AI%20Case%20Study&text=The%20bank%20implemented%20an%20AI,receipts%2C%20and%20other%20financial%20documents.
- *Revolutionizing healthcare: the role of artificial intelligence in ..* https://bmcmededuc.biomedcentral.com/articles/10.1186/s12909-023-04698-z
- *How AI Is Shaping Predictive Analytics in Finance* https://blog.workday.com/en-us/how-ai-is-shaping-predictive-analytics-in-finance.html
- *How AI Is Transforming Inventory Management in Retail* .. https://pavion.com/resource/how-ai-is-transforming-inventory-management-in-retail-operations/
- *How is AI being used in Manufacturing* https://www.ibm.com/think/topics/ai-in-manufacturing
- *How Two Big-Box Retailers Are Using AI* https://www.convenience.org/Media/Daily/2024/June/21/1-How-Two-Big-Box-Retailers-Are-Using-AI_Tech
- *Predictive analytics: 4 success stories - CIO* https://www.cio.com/article/188922/predictive-analytics-4-success-stories.html
- *Revolutionizing Logistics: Case Studies on Successful AI* .. https://www.freightamigo.com/blog/revolutionizing-logistic

s-case-studies-on-successful-ai-integration/

- *How Agencies Are Using AI To Better Personalize Clients ...* https://www.forbes.com/councils/forbesagencycouncil/2024/03/14/how-agencies-are-using-ai-to-better-personalize-clients-marketing/
- *ETHICALLY ALIGNED DESIGN* https://standards.ieee.org/wp-content/uploads/import/documents/other/ead1e.pdf
- *The Intersection of GDPR and AI and 6 Compliance Best ..* https://www.exabeam.com/explainers/gdpr-compliance/the-intersection-of-gdpr-and-ai-and-6-compliance-best-practices/
- *Explainable AI: Enhancing AI Transparency and Trust* https://www.tredence.com/blog/navigating-ai-transparency-evaluating-explainable-ai-systems-for-reliable-and-transparent-ai
- *What is AI bias? Causes, effects, and mitigation strategies* https://www.sap.com/resources/what-is-ai-bias
- *20 AI Tools To Supercharge Your Business And Productivity* https://www.forbes.com/sites/jodiecook/2023/09/11/20-ai-tools-to-supercharge-your-business-and-productivity/
- *Artificial Intelligence | NSF - National Science Foundation* https://www.nsf.gov/focus-areas/artificial-intelligence
- *Change Management for Artificial Intelligence Adoption* https://www.boozallen.com/insights/ai-research/change-management-for-artificial-intelligence-adoption.html
- *AI Performance Metrics: The Science & Art of Measuring*

AI https://www.version1.com/en-us/blog/ai-performance-metrics-the-science-and-art-of-measuring-ai/

- *20 AI Tools To Supercharge Your Business And Productivity* https://www.forbes.com/sites/jodiecook/2023/09/11/20-ai-tools-to-supercharge-your-business-and-productivity/
- *How AI Will Transform Project Management* https://hbr.org/2023/02/how-ai-will-transform-project-management
- *The Role of AI in Data Analytics: Transforming Data into ...* https://www.dataideology.com/ai-in-data-analytics/
- *AI in CRM (Customer Relationship Management)* https://www.ibm.com/think/topics/ai-crm
- *The State of AI: Global survey* https://www.mckinsey.com/capabilities/quantumblack/our-insights/the-state-of-ai
- *How blockchain adds trust to AI and IoT* https://www.ibm.com/think/topics/blockchain-for-trustworthy-ai
- *AI, automation, and the future of work: Ten things to solve for* https://www.mckinsey.com/featured-insights/future-of-work/ai-automation-and-the-future-of-work-ten-things-to-solve-for
- *Emerging Trends In AI In Biotech* https://www.forbes.com/councils/forbesbusinesscouncil/2024/06/24/emerging-trends-in-ai-in-biotech/
- *Build Scalable AI for Real-World Businesses Today - RTS Labs* https://rtslabs.com/build-scalable-ai#:~:text=Cloud%20platforms%20are%20ideal%20for,deploy%2C%20and%20manage%20

AI%20models.

- *What Is Modular AI Architecture? - Magai* https://magai.co/what-is-modular-ai-architecture/#:~:text=Modular%20AI%20architecture%20allows%20organizations,while%20keeping%20the%20system%20stable.

- *Driving AI Success by Engaging a Cross-Functional Team* https://www.datarobot.com/blog/driving-ai-success-by-engaging-a-cross-functional-team/

- *Simplifying AI in Agile Project Management for Success* https://www.invensislearning.com/blog/using-agile-in-ai-and-machine-learning-projects/#:~:text=Agile%20methodologies%20are%20a%20set,and%20deliver%20high%2Dquality%20models.

- *Top 30 AI-Powered Online Learning Platforms for Business* .. https://www.disco.co/blog/top-30-online-learning-platforms-for-business-enhance-your-corporate-training-in-2024

- *4 Ways AI is Transforming Community Learning Experience* https://www.grazitti.com/resource/articles/4-ways-ai-powered-online-learning-communities-are-transforming-learning-experience/

- *Cross-Functional Teams for AI Success: Guide* https://dialzara.com/blog/cross-functional-teams-for-ai-success-guide/

- *The State of AI: Global survey* https://www.mckinsey.com/capabilities/quantumblack/our-insights/the-state-of-ai

About the Author

Roger Best

MEET ROGER BEST

Roger Best is a seasoned entrepreneur, author, and coach with over four decades of experience building and leading successful businesses across multiple industries. From his first venture at the age of eight, digging and selling worms to a local bait shop, Roger has been driven by a relentless curiosity, a sharp entrepreneurial mind, and a desire to create meaningful success, not just financial, but personal and lasting.

After years of pushing through 60+ hour workweeks, skipped vacations, and the ever-elusive promise of "one day," Roger experienced what so many entrepreneurs do: burnout. But rather than accept it as the cost of doing business, he chose a different path, one that led to balance, freedom, and the ability to build a life he didn't need to escape from.

Today, Roger lives and works in Puerto Rico with his wife of more than 46 years, where he runs multiple businesses, writes, coaches, and enjoys the sunshine and laid-back island vibes. His books have helped countless entrepreneurs trade exhaustion for intentional living, and reclaim the freedom they originally set out to find.

Roger's mission is simple but powerful: to help entrepreneurs escape the grind, rediscover their purpose, and design businesses that support the life they actually want. Through personal storytelling, practical strategies, and a healthy dose of humor, his writing resonates with overworked professionals who know there must be a better way and are ready to find it.

Whether through his books, one-on-one coaching, webinars, or mastermind groups, Roger is your guide to building a business that works for you, not the other way around. You'll find me online at RogerGBest.com.

Also by Roger Best

EXPLORE A COLLECTION OF inspiring and practical reads designed to empower entrepreneurs and business owners. From reclaiming your time to building a business that truly serves you, these books offer actionable insights to help you create a successful and meaningful life and business.

From Hamster Wheel to Hammock: A Guide to Taking Your Day Back*(Available on Amazon – ebook, paperback and Audible)*

Escape the relentless grind and rediscover the freedom you dreamed of when you started your business. This practical and empowering guide offers actionable strategies to help you break free from the hamster wheel, reclaim your time, and build a balanced, fulfilling life. Whether you're overwhelmed by long hours or stuck in the cycle of constant hustle, this book will show you the path to a life of purpose and peace.

Personal Growth for Entrepreneurs: Your Time, Your Way*(Available on Amazon – ebook and paperback)*

Take a deeper dive into aligning your business and life with your values and passions. This book focuses on developing habits, mindsets, and systems that not only support your professional goals but also nurture your personal growth. Learn how to create meaningful success that feels true to who you are while finding fulfillment in every step of the journey.

The Liberated Entrepreneur: Building a Business That Works for You (Available on Amazon – ebook and paperback)

Running a business doesn't have to mean sacrificing your life. In this transformative guide, you'll learn how to build a business that aligns with your values, supports your dreams, and works *for* you—not the other way around. Discover strategies for streamlining operations, empowering your team, and creating a sustainable work-life balance. This book is for entrepreneurs ready to escape burnout and reclaim the joy and freedom they set out to achieve.

Purpose-Driven Entrepreneurship: Reclaim Your Why for Lasting Success *(Available on Amazon – ebook and paperback)*

Discover the power of reconnecting with your purpose and unlock the clarity and passion needed to thrive. This book offers practical exercises, journaling prompts, and reflective tools to help you rediscover your core motivations, make aligned decisions, and build a life and business that truly matter. Perfect as a companion to *From Hamster Wheel to Hammock* and *Personal Growth for Entrepreneurs*, this book is essential for anyone seeking deeper connection, fulfillment, and long-term success.

Protecting Your Business in the Digital Age: A Non-Technical Guide to Cybersecurity for Small Business Owners (Available on Amazon – ebook, paperback, and Audible)

Cybersecurity doesn't have to be overwhelming or overly technical. In this straightforward guide, you'll learn how to protect your business from digital threats, safeguard your financial resources, and ensure your customer data stays secure. Designed specifically for small business owners, this book breaks down the essentials of cybersecurity into practical, easy-to-understand steps that anyone can implement. Whether you're a tech novice or simply want peace of mind, this guide will help you secure your business and its future.

Beginner's Guide to Artifical Intelligence: Understanding AI Essentials, Mastering AI Prompts, & Future-Proofing Your Career & Business (Available on Amazon – ebook and paperback)

This Beginner's Guide to AI is your down-to-earth, no-jargon-required entry point into the fascination work of artificial intelligence. Whether you're curious about how AI works, eager to master the air of prompt engineering, or wondering how this rapidly evolving technology will impact your job or business, this book has you covered. Packed with real-world examples, clear explanations, and practical steps, it turns intimidating tech talk into actionable insight. From smart productivity hacks to monetizing your AI skills, this guide helps you confidently navigate the digital age, without needing a computer science degree or a crystal ball.

Each of these books offers unique insights into different aspects of entrepreneurial life. Whether you're seeking to break free from burnout, deepen your personal growth, reconnect with your purpose, or protect your business in a rapidly changing digital landscape, these reads are your guide to a more intentional, successful, and resilient journey. Explore them today and take the next step toward the life and business you've always envisioned!

Made in the USA
Monee, IL
09 November 2026

67381879R00082